ENERGY SCIENCE, ENGINEERING AND TECHNOLOGY

WHAT IS TS_B?

ENERGY SCIENCE, ENGINEERING AND TECHNOLOGY

Additional books and e-books in this series can be found on Nova's website under the Series tab.

ENERGY SCIENCE, ENGINEERING AND TECHNOLOGY

WHAT IS TS_B?

WEIMIN HUANG

nova science publishers
New York

Library of Congress Cataloging-in-Publication Data

Names: Huang, Weimin (Professor of science and technology), author. Title: What is TSB? / Weimin Huang. Description: New York : Nova Science Publishers, Inc., 2020. | Series: Energy science, engineering and technology | Includes bibliographical references and index. | Summary: "This book is intended to answer the following questions: 1. What is ? 2. Why is the momentum order? 3. When and where does irreversibility occur? 4. How can the matter become more orderly? In a relatively well-understood field of my research, the new concepts will be used to explain some of the problems that have puzzled people for about 150 years. The is a type of energy that is missing, not the heat extracted from known energy. It is the potential energy of the spatial configuration of the momentum at the location of the material in the system constructed under known forces. There are two potential energies in the mechanics system. Boltzmann entropy is to calculate the total number of microscopic states by the number of momentum counted in the statistical system. So it must be the order of momentum, instead of the spatial position arrangement of the witnessed substances. The entropy of the ideal gas in the wall-less space can be proved to be equal to considering spatial degeneracy due to space distinguishable of molecule momentum. The gas state equation hides the principle of interaction invariance . One force has two kinds of potential energy. The certainty of the statistical system stems from the order of momentum. The Clausius statement of the second law of thermodynamics is that heat spontaneously flows from a high temperature object to a low temperature one. The Boltzmann entropy of this process is decreased, which is in contrast to the increase of Clausius entropy of the process. This shows that there are two kinds of entropy and two kinds of entropy criteria. The world around us often becomes more orderly. The vortex in the fluid is due to the combination of the angular momentum number and the order flow structure generated spontaneously. When the electrons in the Hydrogen atom form a covalent bond of the molecule, the degeneracy of the electron microscopic states is caused by quantum entanglement. This merge is not only a spontaneous process, but also the molecule is a more ordered material state. Since the mixing process is also a spontaneous process of ΔSB reduction, the mixture is in a more ordered state. These help us understand that a large number of high molecular compounds are crowded into cells to construct highly ordered life activities. Of course, the orderly process is conditional. We can expect to discover more in the application of the second law of thermodynamics"-- Provided by publisher.
Identifiers: LCCN 2020020536 (print) | LCCN 2020020537 (ebook) | ISBN 9781536180039 | ISBN 9781536180046 (adobe pdf)
Subjects: LCSH: Entropy. | Statistical mechanics. Classification: LCC QC318.E57 H83 2020 (print) | LCC QC318.E57 (ebook) | DDC 536/.73--dc23
LC record available at https://lccn.loc.gov/2020020536. LC ebook record available at https://lccn.loc.gov/2020020537

Published by Nova Science Publishers, Inc. † New York

To my beloved wife Peiping Zhang

Contents

PREFACE

During the Industrial Revolution, the steam engine gave birth to thermodynamics and our contemporary knowledge of work and heat. Due to advances in mechanics, we have a pretty clear understanding of how thermodynamics works. On the other hand, the many problems within thermodynamics are closely related to "heat." Because heat is everywhere, the mechanics have never been perfected.

According to a question posited by Maxwell's demon: within distinguishable molecular momentum, is there a spatial divisibility? This book aims to clarify the concept of "heat" in four ways:

1. Explain clearly that p in ideal gas is a force with potential energy;
2. Express a method for thermodynamic statistics where space divisibility is proposed;
3. Show that there are two kinds of potential energies for a force corresponding to a kind of momentum;
4. Discuss how there are two kinds of entropy.

While my aim is to clearly explain concepts that remained unsolved for a long time, I hope I can also extrapolate on the question: what is TS_B? I think it is a cosmic ability to arrange itself.

Perhaps, because I did not recognize TS_B before, I came to realize that it is very important, and I had an urge to write this book. I find that although everyone regards it to be "true," it is also just common sense.

I hope physicists and mathematicians can learn from my research, which can be used in more fields. Although I am not an expert, I write from the perspective of an experienced engineer, and I desire to make more people understand TS_B.

Thank you to the Chinese Nature Science Fund Commission for supporting my work for many years!

Chapter 1

NEW IDEA AND CONCEPTS

1.1. PHYSICAL THOUGHT IN MAXWELL'S DEMON PROBLEM

Maxwell not only established the theory of electromagnetic field, but also made great contributions to thermodynamics. His law of molecular velocity distribution is the cornerstone of modern thermodynamics. However, Maxwell's demon, which he proposed more than 150 years ago, is still inspiring people to explore the unsatisfactory aspects of thermodynamics.

In 1867, he disagreed with Clausius's theory of heat death and proposed Maxwell's demon idea for the first time. Maxwell not only questioned whether the second law of thermodynamics was applicable to open systems such as the universe. It is also based on his understanding of the determinacy of the statistical system. The phenomenological physical quantity on the macro scale is based on the statistical determinacy of the microscopic physical quantity, and this determinacy is not only the statistical average in the mathematical sense, but is caused by the laws of the physical world itself! In 1871 he wrote in the *Theory of Heat*: "One of the most reliable facts determined by thermodynamics is that the temperature and pressure of a system enclosed in a constant volume and non-heat conducting enclosure are maintained everywhere. If no work is done, it is impossible to produce non-uniformity of temperature or pressure, which is the second law of thermodynamics. If we only study large matter and cannot identify or deal with the individual molecules that make up it, this is undoubtedly correct. But if we imagine a creature, it is sensitive enough to track molecules in every movement. This creature, although its ability is still limited as ours, will be able to do things we cannot do. Because we have noticed that the speed of molecules moving in a vessel filled with air in an isothermal state is not uniform, although the average speed of a large number of molecules is uniform. Now we imagine that the container is divided into two parts, A and B, leaving a small hole in the boundary wall-a hole that can be opened and closed by a creature that can see a single molecule, only allowing fast molecules to enter from A to B, and slow molecules to enter from B to A. In this way, it does not need to do work to raise the temperature of B and lower the temperature of A, which contradicts the second law of thermodynamics."

What is worth paying attention to in his statement is: 1. He feels that there is the possibility of such separation in nature, because a small probability does not mean that there is no possibility; 2. The large probability or small probability events are all dominated by internal laws. The "demon" he hopes to find is a certain law. 3. Momentum is separated into two parts. In fact, a new concept of space distribution or "configuration" of momentum is put forward. In other words, is the distinguishable momentum of particles is spatial distinguishable?

The spatial configuration of momentum is the core of Maxwell's demon problem. This concept is rooted in Maxwell's understanding of the "field" of motion formed by concepts of electromagnetism and space-time. Molecules do not move in absolute space-time framework, but under the interaction of elastic collision, the moving molecules will be distributed in space according to certain rules based on their speed, or they will be called the space configuration of momentum. It is different from the spatial distribution of distance formed by the positions occupied by molecules or atoms in space in solids. There also are, of course, position spatial configuration in gas and liquid.

This configuration is constantly changing and cannot be a uniform state. The "fluctuation" that was later accepted by everyone was this idea. But for gas, is there no other momentum configuration except uniform distribution? Is this stable homogeneous state a representation of chaotic molecular motion? Maxwell's deep tone white light is a mixture of seven colors. Can we find a "triangular prism" that decomposes molecular motion? Finding a way to separate molecules by momentum is the first step in Newton's analytical interpretation method, which is the real implication of Maxwell's demon problem.

The second important new concept is "order". Two kinds of different items a, and b are put into A and B respectively, and a and b are mixed put into A and B. We can understand that separate putting will be more orderly. Of course, we also pay more attention to whether a, and b are placed neatly in A and B respectively. In short, it can be called the "order" of positions. When these two parts contain molecules with two different velocities, they are of course also a kind of "order", which is the core of Maxwell's demon problem. We will discuss the "order" of this momentum in space. The "order" of momentum shall be the real world's order. Because everything in the world is moving, order is the "order" of movement.

The third important concept is what "demon" is. Information is introduced from probabilities, thus Boltzmann entropy is introduced into informatics, and "information entropy" is obtained. Demons are also understood as methods and means that do not do work but can obtain system motion information. People have done many ideological experiments and real experiments in an attempt to find Maxwell's imaginary "creatures". As a physicist who has profound physical thoughts and is proficient in mathematics, he must be thinking about a law of the world that has not yet been found, just like Newton's "first driving force", which is just an excuse for "unknown". He hopes that the microscopic world will also follow the second law of thermodynamics. The universe shall be unified. He hopes that the second law of thermodynamics is not only macro deterministic, but also micro deterministic. If the second law of thermodynamics is not established at the level of micro molecular motion, there shall also be reasons for non-conformity, whether positive or negative, which is the "demon" Maxwell is looking for. The speed of the molecule itself is information, and the "demon" recognizes the speed just like the visual perception of color, which is not light itself. Because momentum, a physical quantity, is a trinity of matter, space and implied time, "momentum field" shall be the best expression of this concept. Momentum itself carries the spatial information. That is to say, it has its own "order", that is, we need to find information $P(q)$ about the position q of momentum P , and it does not require any accompanying information.

It can be summarized that Maxwell found the distribution of molecular motion velocity, which is the distribution according to the magnitude of velocity. He hoped to find the spatial distribution information of different molecular motion velocities, that is, the spatial configuration of momentum, that is, the spatial distinguishable of molecules with different velocities. This configuration must be dominated by an internal law in the statistical system. This shall be the Maxwell "demon" we are trying to find. Maxwell's demon and the moon orbiting the earth thrown from Newton's mountaintop are both thought experiments, looking for the great "beauty" of universal laws, and are by no means ugly goblins. It shall be some determinacy.

1.2. No Wall Space Model of Ideal Gas

All thermodynamic discussions begin with the ideal gas model in the box. But in the world that people can see, more gas does not exist in the space with walls. Not only are we in the earth's atmosphere, but also gaseous planets like Jupiter.

All condensed matter is constructed by molecules, which makes no wall space model of ideal gas an important starting point for condensed matter physics.

A white light under the prism makes us understand the colorful world. Maybe this model will bring us the same surprise?

1.2.1. Pressure

The pressure of ideal gas is usually defined by the impulse of molecules hitting the container wall. The problem is that the pressure is a vector. In a square regular hexahedron container, the force acting on the ideal gas only appears in the normal direction of six surfaces, which can be balanced, but cannot be well defined in any other direction. Therefore, this pressure shall be defined on the sphere. The ideal gas in a no wall space must have a spherical spatial distribution so that the definition of pressure to have a rotational symmetry.

In the no wall space, without the wall impulse, the pressure can only be defined by fully elastic collision used by Maxwell in deriving the velocity distribution law of molecular motion. The pressure in the direction of the internal normal at any point in space can only be defined by fully elastic collision of molecules.

The pressure, a generalized binding force defined by fully elastic collisions, needs to be strictly defined. Firstly, because it has not been strictly defined in fact; secondly, there is no place for it in the "standard model" of physics. Although we know that mass is obtained by the Higgs mechanism and the weak forces mediated by the gauge boson with mass is involved in the evolution of many elementary particles, the quarks construct basic particles with the aid of the gluons under strong force, and we also know that Coulomb force and electromagnetic forces construct atoms and molecules, we seem to take it for granted that molecules themselves construct condensed states in the real world, although it is the most basic form in the real world we face in our life, including human beings themselves. The force of condensed state seems to have been found by Van der Waals force, but there is no force in ideal gas. Physics is definitely not a complete self-consistent system, which makes condensed matter physics more like an experimental physics with no source of origin. It is not even clear why the phase transition of gas, liquid and solid occurs. Therefore, the relationship between Van der Waals force and pressure needs to be discussed in detail in the following condensed matter physics problems.

The N ideal gas molecules with uniform density occupy no wall space with volume of V and rely on elastic collision to form rotational symmetry which is only possible if the sphere is distributed. The collision density in the radius direction per unit time per unit spherical area is

$$n_i(T,N)=\frac{N}{V}\sqrt{\frac{kT}{2\pi m}}=\frac{1}{4}\frac{N}{V}\bar{v}\,, \tag{1.1}$$

Where, $\bar{v}$ is the average speed. However, on the integral microelement within the solid angle, the momentum variation difference must be generated on the radial spherical difference $dA=2rdr\sin\theta\cos\phi$ to ensure that the difference in particles number between the inner and the outer spherical surfaces causes the impulse caused by the difference of momentum naturally pointing to the normal direction of the spherical surface. The impulse law is:

$$\int_0^{\Delta t}\sum_j n_j(N_i,T)\vec{f}_j dt=\iiint_V \frac{1}{V}\sum_j n_j(N_i,T)\Delta(m_i\vec{u}_i)_j\, r^2\sin\theta drd\theta d\varphi \tag{1.2}$$

Where, there are N ideal gas molecules in the integral volume V, and the n_j is the collision probability at the temperature T, the momentum difference of each collision is $\Delta(m_i\vec{u}_i)$, $\vec{f}_j dt$ is the impulse generated by these collisions. Due to the symmetry of molecular density in the directions of θ and Φ, in the spherical shell integral microelement, there is no residual momentum difference in θ and ϕ coordinates, so the ideal gas collisions effect only occurs in the radial direction. This is the dynamic reason of rotational symmetry distribution. Therefore, there is only impulse in the normal direction r. The limit of these impulses in a sufficiently short time can be defined as the pressure in the ideal gas to point into the system,

$$p = \lim_{dt \to 0} \int_0^{\Delta t} \sum_j n_i (N_i, T) \vec{f}_j dt = \lim_{V \to 0} \iiint_V \sum_j \frac{n_j (N_j, T)}{V} \Delta (m_i \vec{u}_i)_{r,j} \tag{1.3}$$

$$\frac{-\vec{r}}{|\vec{r}|} r^2 dr \sin\theta d\theta d\phi$$

This force is neither any force in the known standard model nor gravity, but a generalized binding force, i.e., the force produced by elastic collisions between molecules. With the enhancement of van der Waals force between molecules, even the force and translational momentum will be locked during the phase transition. However, collisions of liquid molecules can still exchange rotational and vibration momentums. It still has the generalized intermolecular binding force that system can define-intermolecular shear force and Hook elastic force. Together with Van der Waals force, they form the stress of continuous medium in macro sense.

1.2.2. Pressure Field Uniformity

However, in this solid angle of integral microelements, if the number density N_j/V is uniform along the radial direction, the pressure p will increase in the spherical diameter direction for the solid angle, with the increase of the spherical area. To satisfy the uniform condition of pressure for a thermodynamic system in equilibrium state, the collision density n_j/V must be gradually reduced along the radial direction. Therefore, a no wall ideal gas model with the position spatial configuration of particles is a constrained system satisfying thermodynamic equilibrium conditions. However, this constraint condition will cause temperature variation in the radial direction, because the temperature is based on the statistical average of molecular motion velocities of number densities. The decrease of the number density will cause the temperature difference between spherical shells with the same velocity distribution.

1.2.3. Temperature Field Uniformity

According to Schrodinger equation, the solution of translation kinetic energy of molecular in a solid-walled box space with stationary sides lengths of a, b and c is as follows:

$$\varepsilon_i = \frac{\hbar^2}{m}\left(\frac{n_x^2}{a^2} + \frac{n_y^2}{b^2} + \frac{n_z^2}{c^2}\right) \tag{1.4}$$

Here n is principal quantum number that is a function of side lengths a, b and c. It means the result of elastic collision that the distance between the two particles reflects the magnitude of momentum. The particles with large energy have larger space and larger principal quantum number. The volume of ideal gas in no wall space is constrained by elastic collision. This solution can also be characterized by tetrahedron composed of 4 particles that is a volume with the smallest number of particles. (The distance space defined by the regular tetrahedron is also discussed below) The microelement volume formed by the distance between adjacent tetrahedrons is also related to their average momentum. Moreover, these volumes of this microelement can press on towards the volume occupied by the whole system, that is, the space of the whole "momentum field" is determined by the momentum of all particles in the system! The number of particles N can be composed of $N-3$ tetrahedrons at some time point. The local temperature can be defined by the statistical average of the kinetic energy of the molecules in several tetrahedrons for local volume space

$$T = \lim_{V \to 0} \frac{1}{V} \frac{2}{3R} \iiint_V \sum_j^m n_i^* \varepsilon_i(V_j) dV \tag{1.5}$$

Such a definition can not only characterize the temperature in the no wall ideal gas model, but also help to understand the intrinsic relation between momentum and its space and what the "momentum field" is. Space is the field of momentum.

If the temperature in the no wall model of ideal gas needs to satisfy the second condition of thermodynamic equilibrium that the system temperature is uniform. Under the above-mentioned radial different number density distributions, it is impossible to keep the temperature uniform by the statistical average of this space distribution definable temperature. So that the unequal number of molecules in the radial direction means that the volumes in the tetrahedron are different, the kinetic energy of the corresponding particles is different, and the momentum is also not equal for the same molecules at different positions. When the number density is not equal in the radial direction, there must be a spatial distribution of different momentum in the radial direction.

By the thermodynamic equilibrium conditions, molecules with larger momentum will automatically accumulate in the spherical shell with larger radius through collision, while molecules with smaller momentum will accumulate in the spherical center region. The molecule momentum of ideal gas in this model would be arranged by size along the radius of the sphere. This kind of momentum distribution in space is possible. This is the momentum space configuration that Maxwell wanted to find. The "demon" here is an elastic collision. The spatial distribution of molecular momentum with different sizes only needs to meet the thermodynamic equilibrium conditions of the pressure and temperature in the whole system. Nor does it violate the macroscopic second law of thermodynamics.

1.2.4. Momentum Space Configuration of Ideal Gas in Box

The ideal gas in the box also has momentum space configuration under thermodynamic equilibrium conditions. This microscopic spatial distribution of momentum must be uniform as our common sense tells us. How can this microscopic uniformity of molecules with different sizes momentum having Gaussian distribution be formed? There is only one possibility, that is, the molecules of each velocity are uniformly distributed in the same volume of the box according to each number of every levels of molecules momentum by solution of Schrodinger equation. This is not only the result of molecular elastic collision, but also the result of collision with the box wall, which constitutes the momentum field of every molecules velocity. The momentum space configuration in this box is even superposition of "momentum fields" of molecules with different velocities. That is, the tetrahedron composed of molecules of each velocity according to their interaction distance is uniformly distributed in the same-box space, and packed up in the same-box space V . This microscopic statistical determinacy purports that there is not the disordered movement of particles in the absolute space of the box, but a uniform and orderly momentum field. Fluctuations are only responses to external disturbances, not an intrinsic feature of thermodynamic equilibrium! The random movement does not cause fluctuation. So far we can at least say that momentum space configuration is an objective physical fact for ideal gas molecules. This constitutes the most important cornerstone of this book's discussion.

The no wall model of ideal gas is not only an ideological experiment with physical reality, but also a real physical fact that has been proved by experiments.

1.2.5. Momentum Field

Maxwell has a special interest in "field". In fact, when dealing with this no wall model of ideal gas, the concept of "momentum field" will inevitably be introduced. It has been explained above that elastic collisions can direct pressure inside the system, but the question is whether the gas molecules with maximum velocity in an outermost layer can be "constrained"? Pressure is the only force to maintain an ideal gas system. The produced molecular motion with the moving distance of particle in space and the delayed time are the nature of the body's motion itself. It is that space-time is associated with molecules. Molecules do not move in an absolute space, the space associated with these moving molecules composes the volume of gas, and the collision frequency of the moving molecules is their respective clocks. Therefore, the space accompanied by the gas molecules with the highest velocity in the outermost layer is the result of collision of ideal gas molecules in no walls space. When it runs too far, the molecular momentum is not large enough, and it will regain enough momentum and return to a large enough space position through collision after falling. It will never be far from where it shall be. The "momentum field" of ideal gas in no wall space can be defined as a space constrained by pressure. It is not necessary to use similar gravity as the "first driving force" to build this model here. Only this concept of "momentum field" can be satisfied in a self-consistent way, i.e., the distance between particles the spatial distribution of molecules momentum on its distance whether their momentum is regarded as continuous or quantum distribution. When the total system kinetic energy shows a statistical average phenomenological temperature rise, the volume of the sphere increases and the maximum curvature of the sphere decreases. Therefore, it is said the mass tells the space-time how to bend rather than momentum does.

1.2.6. Momentum Space Distribution of Ideal Gas in Bose-Einstein Condensed State Experiment

The momentum space configuration in the no wall model of ideal gas can be easily verified by numerical simulation of molecular dynamics model. However, perhaps it was the guidance of some unseen deity that actually made me easily consult the BEC experiments made with alkali metal rubidium atomic gas at extremely low temperature published by physicists Nair and Weiman of the University of Colorado and the National Bureau of Standards (who shared the 2001 Nobel Prize in Physics with Professor Chetri of MIT) in the vast ocean of unfamiliar physical documents. These metal atoms are in an ideal gas state. They are bound by magnetic traps without walls. They are slowed down by laser beams and cooled by evaporation (only allowing molecules with large momentum to escape) to a temperature of tens of nK near absolute zero. They measured the spatial distribution of molecular momentum and density by neutron scattering experiments, confirming the existence of atoms with zero momentum in the momentum space predicted by Bose and Einstein. What makes them confused is that the molecular momentum increases from small to large and the density decreases from large to small in the radial direction, regardless of with or without magnetic trap constraint, just as I predicted in the no wall model of ideal gas. They won the prize for this condensed state, but there is no answer to the problem of momentum space configuration. Italian scholars once tried to explain this "strange" phenomenon with dynamics theory. The paper is very lengthy, but it is not said clear. It is not a dynamic problem, but a very simple thermodynamic equilibrium problem. It seems that the physical world can only be illustrated by "simplicity".

This experiment is the most direct experimental evidence for momentum space and is very important! Einstein, who was the first to put forward the concept of "momentum field", has incomparable philosophical thinking in physics.

Such a spatial distribution of molecular momentum may even occur in the distribution of macroparticles. The rainbow is formed by refraction of sunlight by micro-droppings in water mist. The color distribution in the rainbow is fixed. This is indicating that there is a fixed spatial distribution of the droplet sizes refracting these colors. The different sized droplets also correspond to their different sized momentum. The rainbow can be viewed as an arc on the longitudinal section of the ball. Macroscopic particles such as water droplets could also be distributed by some spatially order according to their momentum. It may be important in the phase transition problem.

Come here, we can be more confident to discuss this myriad world on the real physical foundation that we can be sure of. The Creator has built a momentum space configuration, and the world is only a "momentum field".

1.3. Why Is Momentum Order?

People always put things in good sequence and regard them as "order" in their sub-consciousness. Because the position of space is real and visible in people's minds, whether it is satellite positioning, identifying the location of your car in Google Maps on your mobile phone, or using fluorescent materials to show the position of atoms in the molecule and obtain the spatial position of each atom in the biomolecule, all this is to enable us to find ways to visualize the position and spatial configuration of an object when our eyesight is beyond reach, even if it is as small as an atom.

However, the discovery of Brownian motion has confirmed that molecular thermal motion is not in "order", but this is a misunderstanding. The volume of pollen is several orders of magnitude larger than molecules, and the asymmetry of pollen shape makes it impossible for molecules to collide with pollen symmetrically and the movement of pollen must be disordered. The disordered movement of pollen cannot prove the disorder of molecular movement. When discussing the minimization of free energy in biophysics, even if the concept of disorder in *TS* is so incompatible with ordered life, the disordered molecular thermal motion has become an unshakable "iron law". It also makes Maxwell's demon speculation so inexplicable. However, the law of molecular velocity distribution has actually revealed the determinacy of molecular microscopic motion.

Since Nord's principle of symmetry unveiled the curtain of "symmetry" in physics, physicists have reviewed the "change in unchanging" in the "Galileo transformation" and have begun to clearly realize that the world is constantly changing, and motion is the most essential thing in the world. Mountains, trees, streets and houses are all in imperceptible movements, and all movements are controlled by simple laws. Einstein founded the theory of relativity. Because there is no absolutely static reference system, relativity has found the invariable expression of all inertial system. The gravity is equivalent to the curved space-time.

The orderliness of momentum spatial distribution in the non-wall space and the space with wall does not look the same, because there are completely different spatial distributions of positions. What is the expression of the order of momentum spatial distribution, or what shall be used to measure the "order" of the momentum field? That is, how can we find an orderly expression with invariance?

1.3.1. Why Is Boltzmann Entropy "Momentum Order"?

The probability in Boltzmann entropy is the probability of counting the possible combinations of molecules of various momentums according to the size of velocity and possible combinations of locations in space. Its essence is how many kinds of possibilities there are in the spatial configuration of momentums. Boltzmann has already answered the above question, so this can be taken as the starting point for discussion.

The question is whether it can change from a statistical probability with a large numerical value to an expression of "order", i.e., is Boltzmann's probability entropy deterministic? If it is determinable, we will undoubtedly find the most suitable physical quantity to measure this "order". We know that two macroscopic thermodynamic function values can be used to obtain the partition function of the number of molecules with different momentum, this shows that Boltzmann entropy is deterministic. In this way, we only need to find the relationship between this determinacy and the spatial distribution of momentum, to find the "order" with invariance in the space with wall and no wall space.

1.3.2. Local Off-Domain Sub-Statistics of Identical Particles for Ideal Gases

In statistical thermodynamics the identity principle of microscopic particles or the indiscernibility of identical particles means that the exchange of any two identical particles does not produce a new quantum microscopic state. In other words, if the number of particles and the total energy of ideal gas do not change in a no-wall space, under the condition of thermodynamic equilibrium, the exchange of particles satisfying identity only occurs in the $r \to r + dr$ spherical shell space, that is, this identity only exists in the local area. It is the local off-domain sub-domain. The identical particles in different positions are no longer indistinguishable, and momentum is spatially distinguishable. They are the locater for whole spatial volume.

Particles in different regions of quantum states with different energy levels are equivalent to with locality and cannot be exchanged in space

In order to avoid infinite velocity in Maxwell velocity distribution, Boltzmann entropy is discussed using discrete quantum momentum. If in each set of energy level distribution, each quantum level ε_i is in the local space region, only the n_i particles are indiscernible delocalized, and the possible statistical combinations on g_i quantum merging states are $g_i^{n_i}/n_i\,!$ quantum states. The total number of microscopic states is usually

$$W = \prod_{i(N,E)} \frac{g_i^{n_i}}{n_i\,!} \tag{1.6}$$

This statistic is uncertain. In order to determine this statistical distribution, it is usually necessary to introduce a partition function specified by macroscopic thermodynamic quantities. Under the condition of the most probable distribution, the particle number distribution is finally determined by the determinacy of the macroscopic physical quantity of the system.

According to Maxwell's velocity distribution law, the velocity distribution is unique for a certain temperature. However, in the distribution of quantum state energy levels, there may be multiple sets of energy level distributions due to the simultaneous determination of quantum number (including the number of degeneracy) and particle number. However, when the no wall space is balance in pressure, the particle number density distribution in space is the position order of particles in system space. This is determined by pressure (pure elastic collision). The total energy determined by a monatomic ideal gas is equivalent to the temperature is restricted. When the temperature in the space is the same everywhere, it is necessary to specify a certain number of particles in the corresponding quantum state energy levels everywhere in the space, so that the macroscopic temperature can be kept the same in each space radius with different particle density numbers, that is, the spatial distribution of different particle numbers with different energy levels. According to formula (1.5), it is conceivable to define the tetrahedral volume at the same temperature. For different momentum corresponding to different quantum numbers, the tetrahedral volume is different, and the final total volume V of plus sum is the only determinacy for adapting the corresponding thermodynamic equilibrium conditions for p and T. The distribution of the number of particles on the corresponding energy level under thermal equilibrium will be unique. In fact, the spatial stipulation of quantum states and their numbers is essentially the partition of the particles number of according to their energy levels. Therefore, the microscopic particle number and total energy parameters and the macroscopic thermodynamic equilibrium pressure and temperature are the deterministic conditions of thermodynamic system, and shall also be the sufficient and necessary conditions for the uniqueness of this energy level distribution and particle number partition.

Statistics are taken of all possible quantum states for the N particles. For momentum there are I off-domains, each of which has n_i particles. But in space, there are I locators with n_i particles for each locator. The statistics for momentum are called as local delocalized (off-domain) sub-statistics.

Boltzmann entropy is recalculated by local off domain sub-statistics. Statistics are made on the possible quantum states of all N particles. The number of the arrangement of the microscopic states of the I locators in the space is increased to $\sum_I \prod_i n_i!$. Due to the momentum indivisibility of particles for I off-domains on same energy level the microstates combination numbers will decrease by $\sum_I \prod_i 1/n_i!$. Therefore, the number of possible combinations of total microscopic states of momentum by used of local off-domain statistics shall be:

$$W = \sum_{(N,E)}^{I} \prod_i \frac{n_i!}{n_i!} \prod_i \frac{g_i^{n_i}}{n_i!} = \prod_i \frac{g_i^{n_i}}{n_i!} \tag{1.7}$$

This is the same as the microscopic state number of classical ideal gas delocalization statistics on a set of energy level distributions. For a system with a specified E and N, satisfying the conditions of force and heat balance is equivalent to adding two constraints, and its energy level combination and corresponding particle number distribution will be unique. The equation is:

$$W = \prod_{i(N,E,p,T)} \frac{g_i^{n_i}}{n_i!} \tag{1.8}$$

Ideal gas particles meet the thermodynamic limit assumption of $\langle N_r \rangle \ll 1$. That is, the average number of particles in the quantum state ε_i is very small compared with the total number of particles to ensure that only one particle occupies each quantum state. But these particles having same momentum in the local space are not distinguishable. This means that these particles are the degeneracy at the same energy level in the local domain. That is $g_i^{n_i} = n_i^{n_i}$. This degeneracy state can be called as "space degeneracy state" that is essentially the same as the concept of "exchange degeneracy of position" in F-D statistics, like be in the B-E statistics that the number of particles is unlimited at the same energy level. From this indistinguishable degeneracy of local domain, the total possible combination number for all energy levels ε_i should be

$$W = \prod_{i(N,E,p,T)} \frac{n_i^{n_i}}{n_i!} \tag{1.9}$$

Using the Stirling approximation formula $N! = N^N e^{-n}$ of factorial, the total number of possible particle combinations of ideal gas particles in a no wall space shall be

$$W = \prod_{i(N,E,p,T)} e^{n_i} \tag{1.10}$$

$$S_B = k \ln W^* = kN\left(V\right) \tag{1.11}$$

This Boltzmann entropy is a fixed value for the determined number of particles and the total energy, and is no longer a probability. S_B, which satisfies the thermodynamic equilibrium condition, is a definite value, but as a microscopic statistic it has determinacy as a macroscopic physical quantity. Because the thermodynamic equilibrium condition is equivalent to the partition function determined by the two thermodynamic parameters, the Boltzmann entropy value can be determined naturally.

The most important thing is whether is this local off-domain sub-statistics also applicable to ideal gas in the space with wall. We hope that the "order" of this momentum is constant.

If we put this relation into the equation of the state of ideal gas, we can get this relation:

$$pV = nRT = S_B T \tag{1.12}$$

For a certain number N of particles, S_B is a state function. Under the condition of thermodynamic equilibrium, it is also a thermodynamic function. It is a function of the momentum field space. Rewritten it $\not{V}$ is to highlight the inclusion of momentum field space configuration. The $\not{V}$ can be defined as the "distance space" between particles generated by elastic collision interaction.

With the help of the process heat the Clausius entropy is also defined as a state function, these processes have the same initial and final states undergo different thermal processes, but the formula (1.12) cannot be obtained. The two entropies are different. The following is a further explanation.

It is hoped to explain that the ideal gas in the wall space still has the above relationship.

1.3.3. Change in Invariability for Momentum Space Configuration

For ideal gas uniformly distributed in a container with wall, the number of molecules for different momentum level is determined according to Maxwell's law of gas velocity distribution. It is the same for system with same number of particles and same temperature. We still use finite quantum state momentum to discuss the order of "momentum field" of ideal gas in the box according to solution (1.4) of Schrodinger equation. The momentum space configuration in which molecules with different momentums have been given in the previous section must be uniformly packed up in the volume V according to each distance space scale specified by the quantum number. It shall be noted that only these tetrahedrons are regular tetrahedrons that is the velocities of the 4 molecules on them are equal. According to formula (1.4), any tetrahedron is equivalent to different a, b and c, and the velocities of the 4 molecules in the tetrahedron are not equal. Such tetrahedrons are not stable when they are packed up. Any type tetrahedron cannot be packed up. In fact, only regular tetrahedrons or regular polygonal bodies can be packed up, just like Russian dolls, they must be geometrically similar to each other. This means that elastic collisions will not change this geometry. There is no momentum exchange for collisions between molecules with the same momentum. The volume of regular tetrahedron of that same size does not change with time. The molecules on the large tetrahedron collide with the small-speed molecules on the small regular tetrahedron, and the small-speed molecules bounce off to the inside. The large-speed molecules on the large tetrahedron will remain on the outside by the momentum exchange with particles exchange. We stand on each momentum tetrahedron and observe the structure of the same momentum molecule. Its configuration is unchanged. This is Galileo transformation at the molecular level. The disordered molecular motion seen in absolute coordinates is completely in order in relative coordinates. It's like watching the water impact the bow on the shore, but on the ship, the water flows smoothly along the bow. Molecule at each energy level has its own spatial configuration and clock. For configurations with different momentums, they all have their own configurations under different clocks. However, because the molecular motion speed is much lower than the speed of light and their spatial size is limited, there is no relativistic effect in continuous media. It can be considered that these occur under the same clock, and elastic collisions between these configurations occur continuously, but the net collision effect does not change the "order" of the overall "uniform" pack. Thus large tetrahedrons contain small tetrahedrons, that is, molecules with large momentum contain molecules with small momentum and molecules with small momentum contain molecules with smaller momentum, layer by layer. This momentum packing up method is the same everywhere in the whole space.

There has an internal impulse, which is the generalized binding force-pressure p generated by the elastic collision of molecules. The packing up of regular polyhedrons will not collapse due to molecular elastic collision. That is to say, this geometric space configuration, or distance space configuration, is stable. This form of momentum distribution makes momentum-distinguishable molecules spatially distinguishable. The regular tetrahedral of each scale is withed the more lager tetrahedral. They don't have to be full, so thus space may be a fractal dimension. This ensures that the even momentum space configuration is still a local off-domain sub-domain and locator for whole space. The molecule microstates of each energy level locally are the indistinguishable degeneracy. Then the formulas (1.9)- (1.11) still set up in box space with wall. This is the invariance of space order for momentum in the variation of different spatial position configuration.

The number of particles distributed according to momentum is distinguishable localizer even though tetrahedrons of different sizes are located in the same space, even though they are packed up. Particles in tetrahedrons of the same size are indistinguishable local off domain statistics, so their Boltzmann entropy is the same as the Boltzmann entropy in the no wall space. This is the invariable change at the level of molecular motion. That is, the spatial distribution forms are different, i.e., the "order" in the particle location space is different, and the density is "uneven" or even; then the "order" of momentum in the space specified by momentum is the same.

From this it can be seen that the position order of the ideal gas system in the no wall space and the box can be changed, which is variation of an invariant. However, the momentum order is an invariant in variation. That is, the combinations number of the microscopic states for a system occupying same volume is constant. In other words, Boltzmann entropy for same system is constant under thermodynamic equilibrium conditions. Therefore, the order of the system can only be the "order" of the "momentum field". This is a more essential concept about "order" of material movement.

For a statistical system such as ideal gas, particles with the same energy level will be uniformly distributed in the container space or in a spherical shell with a certain radius in the no wall space. Although the spatial configuration of "momentum field" is completely different, it is determined that randomness is only interchangeability of a particle in this specified spatial region, so there is no concept of the most probable state. No molecules will all appear in the corners of the container, while the rest of the space will be empty. However, in the no wall space, the molecules with large momentum will always be in the outer layer, the molecules with small momentum will always be in the center of the sphere, and wrong positions are not in the statistics. The tactical discipline of these elements to follow the instructions of the coach is extremely "strict", just like goalkeepers are never allowed to run out of the forbidden zone. This is the same as the classical Boltzmann entropy calculation, which can only be determined after the partition function of the molecule is determined by two phenomenological physical quantities. Thermodynamic equilibrium conditions make Boltzmann entropy no longer a probability, but a certain value. It makes the "order" of the momentum field measurable, because the most probable combinations in space are the same, making any specific momentum field order invariant.

This is the symmetry in the microscopic system under the macroscopic determinacy of thermodynamics.

1.4. Two Kinds of Potential Energy

So far, physical chemistry course had all said, “there is no interaction force in ideal gas, so ideal gas has no potential energy.” But how is this self-consistent with the free energy pV in thermodynamics? If the pressure p is still regarded as the force balancing the external force on the cylinder piston and S_C is the state quantity calculated by heat, then the thermodynamic system has only internal energy, and these concepts are very confusing and incomplete. Enthalpy and free energy are proposed as functions, which mean that they are a kind of energy in the system. Thus, there is no longer any heat and work between the system and the environment. Starting from the free energy, understanding the two energies pV and TS is the right direction. It is also a way to extend this symmetry of $pV = TS_B$ from ideal gas to all thermodynamic systems.

1.4.1. Pressure and Two Potential Energies in Ideal Gas

In the (1.10) obtained from the state equation of gas, whether pV and TS_B has potential energies can be checked according to the definition of analytical mechanics. For a single-valued differentiable scalar function $V(x_i, y_i, z_i)$, the force component F_x, F_y, F_z acting on the macroscopic particle of the system has the following relation

$$F_x = -\frac{\partial V}{\partial x_i}$$
$$F_y = -\frac{\partial V}{\partial y_i} \quad (i = 1, 2, ...N) \tag{1.13}$$
$$F_z = -\frac{\partial V}{\partial z_i}$$

Then the system is a conservative force system and the function $V(x_i, y_i, z_i)$ is called system potential energy.

For the spatial distance distribution $\not{V}(r, \theta, \phi)$ of molecules in the model of ideal gas without walls (marked with $\not{V}$ is to distinguish potential function $V(x_i, y_i, z_i)$ and to distinguish geometric space volume), at the $V = V(p\not{V}(r))$ thermal equilibrium state, p is the constant, then for N particles, the distance distribution between them is as follows:

$$-p = -\frac{\partial(p\not{V}_i)}{\partial r} = -p\frac{d\not{V}_i}{dr}$$
$$\frac{d\not{V}_i}{dr} = C = 1 \qquad (i = 1, 2, \cdots N) \tag{1.14}$$
$$\not{V}_i = \not{V}(0) + ar$$

Therefore, it is a potential energy, and the distance space size of molecules has a distribution along the radial direction, which is also the distribution of particle number density. The $p\not{V}$ potential energy is linearly distributed along the radial direction in a spherically symmetric space without walls.

In a wall box space, for the above formula, $\frac{\partial(p\not{V})}{\partial x_i} = \frac{\partial(p\not{V})}{\partial y_i} = \frac{\partial(p\not{V})}{\partial z_i} = 0$, where $p\not{V}$ is a constant, indicating that in a wall box space, the size distribution of the molecular distance space is uniform, and in a wall box space, the potential energy is uniform everywhere.

The relation $p\not{V} = TS_B$ is not only quantitative one, but also indicates TS_B finally is still a function of distance space. In a no wall space, the S_B can be marked as $S_B\left(\not{V}_i\left(r\left(n_j\right)\right)\right)$ for n_j particles of each energy level, so that this configuration of particle number in space according momentum size must also be a potential energy. According to the above definition of potential energy in the sense of dynamics, it can be proved that TS_B is also the potential energy of ideal gas. Their corresponding systems have same $\not{V}_i$ that the spatial distribution of momentum and then changes as $\not{V}_i$ varying

By this definition, $p\not{V}$ is a local description under the meaning in dynamics. In other words, (1.12) can also be applied to the four state functions in all process.

Yet p is a state quantity that is completely independent of the external force. Different from pV in the original free energy, $p\not{V}$ has no concept of work, but the potential energy of the system itself.

The relation $p\not{V} = TS_B$ points to the determinacy between the two spatial distributions. It is also the determinacy between the two energies. If this concept is universal, the two potential energies are not only expressed by microscopic states of molecular motion of ideal gas. It is of great significance to the whole thermodynamics.

1.4.2. Symmetry in Maxwell's Thermodynamic Relations

Under isothermal and isobaric conditions, they are included in Gibbs free energy and begin to appear in macroscopic thermodynamic functions, they should no longer have anything to do with work and heat. But if they are not defined in momentum field, they still can't get away from the concepts of heat and work. So this replacement can return these two potential energies to the energy of the system itself.

Because various "heat" are included in free energy, thermodynamic systems involve more extensive interactions. Free energy can be used in thermodynamic systems with heat and mass exchange by the statistical concept of grand canonical ensemble. Besides heat exchange, heat also involves phase change heat, reaction heat, mixing heat and heat of solution and so on another kinds of heat. These processes also involve changes in system mass (including changes in system composition caused by mixing process); As far as chemical reactions are concerned, the Coulomb force and electromagnetic interaction is only involved. Free energy will not involve intermolecular pressure, Van der Waals force, and Hooke force, only extend from gas gathering state to liquid and solid condensed state. Free energy actually extends from ideal gas to a much wider thermodynamic system. The proof of the more general thermodynamic relationship is to hope that thermodynamics can be extended to a wider range of interactions. Therefore, if the relation $p\not{V} = TS_B$ obtained from ideal gas is proved to be universal in Maxwell's thermodynamic relation, it can be proved to be applicable to a wider thermodynamic category.

Maxwell thermodynamic relation is obtained from state functions of the energy, enthalpy and two kinds of free energy defined by total differential. It is applicable to all thermodynamic systems that can define these thermodynamic functions. Therefore, establishing the connection between Maxwell's thermodynamic relation and theorem $p\not{V} = TS_B$ can prove that it is applicable to all thermodynamic systems.

The original first law of thermodynamics and the expressions of enthalpy and free energy, as well as Maxwell's thermodynamic relations, are used and rewritten by $p\not{V}$ and TS_B as follows

$$
\begin{aligned}
dU &= TdS - pdV & dU &= -TdS_B - pd\not{V} \\
dH &= TdS + Vdp & dH &= -TdS_B + \not{V}dp \\
dA &= -SdT - pdV & dA &= -S_B dT - pd\not{V} \\
dG &= -SdT + Vdp & dG &= -S_B dT + \not{V}dp
\end{aligned}
\tag{1.15}
$$

$$
\begin{aligned}
\left(\frac{\partial T}{\partial V}\right)_S &= -\left(\frac{\partial p}{\partial S}\right)_V & \left(\frac{\partial T}{\partial \not{V}}\right)_{S_B} &= \left(\frac{\partial p}{\partial S_B}\right)_V \\
\left(\frac{\partial T}{\partial p}\right)_S &= \left(\frac{\partial V}{\partial S}\right)_p & \left(\frac{\partial T}{\partial p}\right)_{S_B} &= -\left(\frac{\partial \not{V}}{\partial S_B}\right)_p \\
\left(\frac{\partial S}{\partial V}\right)_T &= \left(\frac{\partial p}{\partial T}\right)_V & -\left(\frac{\partial S_B}{\partial \not{V}}\right)_T &= \left(\frac{\partial p}{\partial T}\right)_V \\
\left(\frac{\partial S}{\partial p}\right)_T &= -\left(\frac{\partial V}{\partial T}\right)_p & \left(\frac{\partial S_B}{\partial p}\right)_T &= \left(\frac{\partial \not{V}}{\partial T}\right)_p
\end{aligned}
\tag{1.16}
$$

In the above relationship, S is replaced with S_B, V is replaced with $\not{V}$. The following will discuss specifically that these two entropies. The Maxwell's thermodynamic relation with the help of Boltzmann entropy can more accurately explain its significance.

In Maxwell's relation, p and T is the intensive quantity, which is for the momentum and energy of the system respectively. The physical quantities $\not{V}$ and S_B is the extensive quantity respectively, which is the partition of the total volume and the total particles number of the "momentum field" for a system. They were once regarded as four independent variables but shall be regarded as a set of related functions characterizing thermodynamic system from a microscopic perspective. It also represents the only four basic processes of thermodynamic system.

Assuming that these four functions have the following relation, the Maxwell relation is used to solve the four undetermined functions.

$$S_B = a(T)p\not{V} \quad p = b(\not{V})TS_B \quad T = c(S_B)p\not{V} \quad \not{V} = d(p)TS_B$$

$$\left(\frac{\partial S_B}{\partial \not{V}}\right)_T = a(T)p = \left(\frac{\partial p}{\partial T}\right)_{\not{V}} = b(\not{V})S_B$$

$$\left(\frac{\partial S_B}{\partial p}\right)_T = a(T)\not{V} = \left(\frac{\partial \not{V}}{\partial T}\right)_p = d(p)S_B$$

$$\left(\frac{\partial T}{\partial \not{V}}\right)_{S_B} = c(S_B)p = \left(\frac{\partial p}{\partial S_B}\right)_{\not{V}} = b(\not{V})T \tag{1.17}$$

$$\left(\frac{\partial T}{\partial p}\right)_{S_B} = c(S_B)\not{V} = \left(\frac{\partial \not{V}}{\partial S_B}\right)_p = d(p)T$$

$$a(T) = \frac{1}{T} \quad b(\not{V}) = \frac{1}{\not{V}} \quad c(S_B) = \frac{1}{S_B} \quad d(p) = \frac{1}{p}$$

This is similar to the process obtaining the gas state equation from the three gas laws which also gets the function relationship $p\not{V} = TS_B$. It is a quantitative relationship so the sign of partial derivative does not be considered.

Maxwell relation is a functional relation among these four variables constructed by these processes. It is equivalent to relation $p\not{V} = TS_B$. Therefore, it is a universal symmetric relation for all thermodynamic systems. In this way, the physical meaning of Maxwell thermodynamic relation can be understood more deeply. Below, we will discuss why replacement of S_C with S_B and V with $\not{V}$ is needed. And why p must be the interaction within the system, not the force between the environment and the system.

The meaning expressed by the two isothermal processes in above relations (1.16) is that to keep the energy of the system unchanged, the change of p and $\not{V}$ means that the closed system is accompanied by a thermal process, and the change of the total energy indicates that the system is not the original system. Only in this way can there be changes of S_B.

Changes of S_B to open systems can be caused by changes of volume $\mathcal{V}$, for example when gases are mixed at equal pressure; Or by changes in force, such as phase change. The most important thing is to realize that the change of the two potential energies cannot be converted by kinetic energy in the isothermal process, nor can they be directly converted by the two potential energies in the system, but through changes in the system itself. The corresponding relation between the two isothermal processes also shows that S_B is a function of $\mathcal{V}$. The two configuration spaces $\mathcal{V}$ and $S_B(\mathcal{V})$ are momentum field and energy field respectively. There is only one acting force to change the two fields, so $S_B(\mathcal{V})$ is a functional. There is no "entropy elasticity". Therefore, using S_B instead of S_C makes free energy no longer has the concept of heat exchange with surroundings.

The two isentropic processes show that only when the force p or the distance space $\mathcal{V}$ changes the energy level distribution in the total energy E, that is, the thermodynamic equilibrium condition changes, can the particle partition in each energy level remain unchanged. It is not that $\Delta S_C = 0$ can be realized without system heat exchange. When the active power of the system is exchanged, such as the compression work changes the potential energy of the original system, the temperature of the system rises, the smaller $\mathcal{V}$ and steeper velocity distribution keep the partition of particle numbers in each momentum constant. The interpretation of energy changes in the system is a process of $\Delta(p\mathcal{V}) = S_B \Delta T$. The adiabatic work changes the total energy of the original system, so the system itself changes and may have $\Delta S_B \neq 0$. The adiabatic work involves exchanging energy with the surroundings, if the free energy need be cut with surroundings that the force must be the interaction defined in the system. The distance space created by this force is not any geometric space.

These explanations are intended to show that the free energy of original associated with work and heat is linking the environment. The only free energy is replaced by new variables that will be the energy in its own system. What the change in the free energy during the process is the conversion between the energy in one's own system. The process reflects the conversion between energies within the system.

To sum up, Maxwell's thermodynamic relation group explains: 1. Energy of thermodynamic system is defined by four interrelated functions, including kinetic energy and two potential energies. 2. Maxwell thermodynamic relations are equivalent to $p\not{V} = TS_B$. 3. The two potential energies are related by the same force. Therefore, it is compatible with the dynamics and conservation of mechanical energy of any known conservative force system. 4. This kind of dependence of $S_B(\not{V})$ determines that there is no mutual transformation between the two potential energies under the same interaction. Therefore, the dynamic equation will not be influenced and can be solved independently. 5. $p\not{V} = TS_B$ is applicable to the range within reach of thermodynamic system.

1.4.3. Hamilton Function

The potential energy TS_B is also consistent with the existing dynamic system, so Hamilton function shall include this potential function. The process is just a dynamic process. Rate of process does not change the nature of dynamics. Using free energy analysis, without exchange, the reversible condition of infinite slowness is redundant. It is a need that there has still thermodynamic relation $p\not{V} = TS_B$ using Hamilton function to explain the dynamic process in.

For a single molecule (particle), whether the classical mechanical method or the Schrodinger equation is used to describe its translational, rotational or vibrational momentum and corresponding energy, if there is no corresponding translational, rotational and vibration potential energy this means that there will be no interaction force with any mechanical description, including Hamilton canonical equation and Liu Wei equation, the force shall be zero. But even if there is no interaction force, the potential energy can be a constant. The idea that there is no force in ideal gas because there is no potential energy in ideal gas is not only inconsistent with the fact that "pressure" is real in the real world, but also contradicts Hamilton canonical equation.

The basic physical quantities to describe Hamilton quantities are geometric space coordinates and momentum of particles, which is an expression of absolute space-time. However, the result of any force is the distance between particles or moving distance. Potential energy refers to the energy corresponding to the "force" of the kind of energy that maintains the relative position between particles. Without distance, there is no potential energy. The potential energy in absolute coordinates shall also be expressed by distance. Galileo transformation or Lorentz transformation of inertial system can ensure that the distance space and momentum in the motion field can always be transformed to the distance space and momentum in absolute space-time. So here we use the distance space in the momentum field to describe it without losing generality.

Potential energy accumulates in the space constructed by the distance between particles, which is essentially a "distance space" rather than a geometric space. In Hamilton, it is possible to use momentum p_i and "distance space element" $\delta \not{V}_i$ as basic physical quantities. This $\delta \not{V}$ is a basic spatial unit formed by the distance between molecules (particles). It can be a three-dimensional volume element, a two-dimensional area element, or a one-dimensional line element.

$$H = H\left(\delta \not{V}_1, \delta \not{V}_2, \cdots \delta \not{V}_N, p_1, p_2, \cdots p_N\right) \tag{1.18}$$

Particles have momentum and energy, and this distance space is the physical space in which their corresponding momentum and energy reside. This space is dynamic because of the hidden time variable, but under the condition of thermodynamic equilibrium, the time-averaged statistical value shall be determined macroscopically. So, Hamilton function can be:

$$H = T + V\left(\delta \not{V}\right) + S_B\left(\delta \not{V}\right) \tag{1.19}$$

Where T represents kinetic energy, V is potential energy of position space configuration and S_B is potential energy of momentum space configuration. If the pressure p can be extended to Van der Waals force and Hooke force, then the potential energy of position space configuration of macroscopic matters for some distance space $\delta \not{V}_i$ can be generically recorded as:

$$V(\delta \not{V}) = p\delta \not{V} \tag{1.20}$$

The macroscopic Hamilton canonical equation for this potential energy of position space configuration shall naturally be:

$$\vec{F} = -\frac{\partial V}{\partial(\delta \not{V})} = -p = \frac{\partial H}{\partial(\delta \not{V})} \tag{1.21}$$

For the position configuration space constructed by the distance space unit, the macroscopic thermodynamic variable temperature corresponding to the total energy also specifies the partition of the total particle number at the energy level. This partition bases on the size of $\delta \not{V}$. From the analysis of classic mechanics, the larger the distance space, the smaller the density and the larger the momentum of particles, so that the system can meet the thermodynamic equilibrium conditions of equal T everywhere. From the perspective of quantum analysis, the size of the distance space unit is exactly the potential well space of kinetic energy of molecules, and the ideal gas is just an aggregation system with enough distance space to satisfy the condition of infinite potential well. Therefore, the larger $\delta \not{V}$ is, the larger the quantum number of the corresponding momentum is, and the higher the energy level is. In other words, the partition of the number of particles in each energy level under the total energy is specified. Therefore, as long as the distance space is measurable for the momentum of a certain particle, the energy partition capability is universally expandable. So we can define

$$S_B(\delta \mathcal{V}) = k\int_j dn_j(\delta \mathcal{V}_i) \qquad (j = 1, 2 \cdots I) \tag{1.22}$$

Since any particle outside a regular tetrahedron can be reconstructed into a new regular tetrahedron. Therefore, it is sufficient to define p by full elastic collision, where N molecules (particles) have N units of $\delta \mathcal{V}$. Therefore, this integral is an integral of the "number" for distance space element, not a volume integral. The definition of this potential energy is for the whole system. It is different from $V(\delta \mathcal{V})$, which can be described locally to the system. The quantum description can be written as:

$$S_B(\delta \mathcal{V}) = k\sum_J n_i(\delta \mathcal{V}_j) \qquad (j = 1, 2 \cdots J) \tag{1.23}$$

There are J energy levels in thermodynamic system, and the number of particles n_i corresponding to each energy level is summed up. The spatial scale of potential well corresponding to each energy level is $\delta \mathcal{V}_j$.

The system p needs to be matched with $\delta \mathcal{V}_j$ corresponding to each energy level of the system, and $S_B(\delta \mathcal{V})$ has the ability to change the number of molecules on mismatched ones. This is why they are called potential energy. Because of symmetry, the two must be equal and only change synchronously. Therefore, the dynamic equation is still the original one, as long as the type of interaction remains the same.

The energy defined by Hamilton function can also define both locally and throughout the momentum field in the dynamic process. Therefore, the relationship $p\mathcal{V} = TS_B$ can also be used in dynamic processes.

1.4.4. Discussion about Distance Space Element δV

Ideal Gas can only be composed of the five regular polyhedrons of regular tetrahedron, regular cube, regular octahedron, regular dodecahedron and regular icosahedron. The regular triangle consists of regular tetrahedron, regular octahedron and regular icosahedron. If there are more than 5 vertices outside each vertex, the spatial configuration by regular triangle will be to degrade into a plane configuration of the triangles. However, if it is an arbitrary tetrahedron, there will be numerous combinations of spatial elements. This arbitrary tetrahedron also means that the spatial combination between any momentums is arbitrary. The pressure defined in local region and related potential energy will be large fluctuate in time and non-uniformity in space. So that there are not the conditions of thermodynamic equilibrium. Yet the regular polyhedron has the optimal (less) combined number of distance space elements by momentum. Moreover, according to the quantum solution of molecular Schrodinger equation, this arbitrary tetrahedron means that four quantum states can appear in this potential well. This will also make the permutation number of molecules to become uncertain at each energy level. The system becomes an indistinguishable off-domain system over the whole space. Therefore, the statistical origin of the distinguishable system as a locator lies in the spatial distinguishability of the distance space unit size. It is δV_i of different scales that make the momentum of particles not interchangeable in space. For the ideal gas in the box space with wall, p and T are uniformly originate from the packing up of molecules with different energy levels, while only the same kind of regular polyhedron can satisfy the requirement of packing up with different sizes .

The solid crystal lattice is regular, not necessarily regular polyhedron, δV_i are identical, but the local vibration momentum is distinguishable, the vibration momentum is not exchangeable in the whole body, and the crystal is still a distinguishable locator subsystem.

The thermodynamic system p defined by δV can be applied to any momentum. The interaction can cause the distance space δV (two-dimensional $\delta\not{A}$ or single-dimensional $\delta\not{L}$) between particles in the system. The particles can be atoms, molecules, clusters of molecules, ions, spin magnetons, electrons and nuclei in atoms and so on. The momentum can be translational, rotational, vibration, spin magnetic momentum, electron momentum in molecules and atoms, and spin momentum of electrons and nuclei, etc. The change of force in the system can transform the distance space $\not{V}$, and there is a corresponding new position potential energy. Only in the system of absolute zero temperature without any momentum can there be no interaction force and have $\delta\not{V} = 0$.

1.4.5. Expansion of TS_B Potential Energy

In view of the generality of Maxwell's thermodynamic relations and the independence of kinetic equations under the invariant condition of interaction, what this book is trying to comprehend action of TS_B in thermodynamic and kinetic processes. From the statistics nature of thermodynamic objects, the quantum state is not an obstacle, so it shall be completely applicable at least at the level of continuous medium and in the scope of Coulomb force and electromagnetic force. For other forces in physics, such as strong interaction and weak interaction and gravity, I have not studied them and will let them go for the time being. If you think this theory has reference value, please leave it to physicists to discuss.

However, in the following discussion, special attention shall be paid to one of the above results: thermodynamic systems can only be discussed under one interaction. Free energy is not a hodgepodge. Correctly defining free energy can correctly interpret the object of the problem.

The absence of TS_B does leave many enigmas in various fields, such as phase transition in condensed matter physic, the infinite propagation speed of heat, the concentrated vortices, tornadoes and typhoons in fluid mechanics and meteorological science, the order of life produced from chaos, etc.

1.4.6. Symmetry $p\not{V} = TS_B$ Dominate Interaction

Mr. Yang Zhen-Ning proposed the principle of “Symmetry Dominate Interaction”. According to this principle, this symmetry dominates that the two potential energies on distance space $\not{V}$ in the system only have one interaction, that is, the force in the system is the same for the two potential energies. It includes the pressure in the ideal gas that is not in standard model, and the Coulomb force and electromagnetic force in the standard model. Because Young-Mills theory applies to other two kinds of interactions except electromagnetic force, therefore, this symmetry may not be limited to electromagnetic force systems.

The invariance of interaction is based on “momentum field”, which corresponds to two kinds of potential energies corresponding to momentum and energy respectively, and they shall have space-time translational symmetry. Because the space and time coordinates all corresponds motion, so that it has invariance of coordinate transformation.

Thus, we can get the most important conclusion. 1. In the thermodynamic equations (1.15) and Maxwell’s thermodynamic relations（1.16), the TS_C can be replaced by TS_B, TS_B is a strictly defined potential energy of the system, and it has a strictly quantitative relationship with $p\not{V}$. Because pV is associated with external force and work, it can be arbitrary and the TS_C calculated from the process thermal effect can be any quantity for a system that will not be with a constant internal energy, and the system will change with heat and work , so enthalpy and free energy are not strictly definable functions for deterministic system. Only TS_B potential energy can be completely cut with thermal effect. 2. The deterministic system has the quantitative relationship $p\not{V} = TS_B$ between the two potential energies by using $p\not{V}$ instead of pV to cut out relation with the work. There can also be a quantitative relationship between kinetic energy and $p\not{V}$ in the following Virial theorem. Such a system is energy with determinacy. But the system corresponding to pV and TS_C is uncertain.

This is in contradiction to thermodynamic statistics specifying the total energy of the system

1.5. Two Entropy and Two Entropy Criteria

1.5.1. The Second Law of Thermodynamics, the Inequity of Work and Heat, Reversible Process

The steam engine was the symbol of the first industrial revolution, and it also opened the history of human understanding the second law of thermodynamics. The equivalent between work and heat proved the equivalence relation in quantity, while Carnot cycle process answers that work and heat are not equivalent. However, the second law of thermodynamics is closely related to "process". It is unscientific to connect the "work" of macro movement with the "heat" of micro movement through the "process". Looking back on the second law of thermodynamics, the only formulation that has nothing to do with the cycle process of the heat engine is Clausius statement: "It is impossible to transfer the energy of a low-temperature body to a high-temperature one in the form of heat without causing other changes." This is equivalent to explaining that the heat never flows spontaneously from a cooler to a hotter body. The second law of thermodynamics is an axiom to interpret the spontaneous direction in the process of microscopic molecular motion, but this requires a form of mathematic expression.

The limit of heat engine efficiency is found by reversible Carnot cycle, but the criterion that is not reversibility itself and is not the second law looking for. The cycle process along any path has the result $dS_C = 0$. The direct physical content is to find a new state function $dS_C = \delta Q/T$ in innumerable processes of conversion between heat and work. The entropy is an invariant for the changes in different processes. The discovery of this entropy is so surprising that people have been stirred in their hearts for nearly 170 years with the mentality of finding the best treasure. Even after

Boltzmann gotten a completely different entropy $S_B = k\ln W^*$ based on molecular thermal motion, people just tried to pull the two together. No one has ever questioned whether the two are the same physical quantity.

But the opposite of the two is doomed from the starting point. The S_B is the real state variable of the system specified by the state number of molecular motion while S_C is not a state variable that can be defined independently, but an increment quantity that can be calculated only by the heat of reversible process. It has no origin. The third law of thermodynamics is used to calculate its absolute value according to the zero value $S_C(0K) = 0$ specified by the experimental results. And this is a variable due to heat transfer, so it cannot be said to be the real state quantity of a system!

Because it is a process in nature, that is the cyclic process on the state space, the inequality for judging reversible process is found in turn.

$$\oint dS_C = \oint \frac{\delta Q_{rev}}{T} \geq 0 \tag{1.24}$$

This inequality cannot prove the causal relationship between the reversibility of the process and the inequality of heat and work demonstrated by Carnot cycle. The process calculating entropy is limited by invertibility. But the loss in system energy conversion can be reduced to zero then the loss of heat and work conversion will still not be reduced at all. Therefore, reversible or irreversible perpetual motion machines of the second kind are impossible to achieve. That is to say, this formula does not show the spontaneous statement of the second law of thermodynamics, nor does it prove that there must be two heat sources to have a heat engine. It should be said that the second law of thermodynamics is self-evident as an axiom. The inequality only shows the difference between reversible and irreversible.

However, formula (1.24) creates a logical chain: molecular motion is disordered, disordered molecular thermal motion loses more work and energy in irreversible process, and the heat that turns out to be is also disordered molecular motion; Irreversibility is the source of disorder; The end of the world is chaotic “heat death “. This irreversible statement has a profound impact. People are told that the room will become more and more chaos if they don’t work hard to clean it up that is entropy increase! The TS in the free energy is that the disordered thermal motion of molecules opposes the orderly mechanical motion, you name it. But (1.24) the criterion of this cyclic process can’t prove the reversibility of any general process except for the cycle process. As a result, there are the difference concepts about entropy production and entropy increase.

The molecular movement itself could be orderly. In the process of radiation and heat conduction, energy transfer is completed in the form of wave, and there is not a lot of energy loss in the wave transmission of heat. The heat is part of deterministic transfer, and irreversibility is caused by momentum of microscopic statistical uncertainty. Irreversibility is a phenomenon when motion momentum at different levels occurs simultaneously. Irreversibility is a random disorder motion included in macro-ordered motion.

Maxwell expressed very clear doubts about Clausius interpretation of eternal entropy increase extended into the theory of “heat death “in the universe. We also heard Einstein’s clear negation of the uncertainty in quantum mechanics: “God does not roll dice”. Through the above systematic discussion on molecular microscopic motion, we have not only found a definite and orderly expression of microscopic statistical systems, but also can find a formal criterion to express the spontaneous tendency of the second law of thermodynamics to any natural processes.

1.5.2. Criterion for Spontaneous Process of the Thermodynamic Second Law

We use Boltzmann entropy to discuss Clausius statement of the thermodynamic second law, that is, heat spontaneously transfer from high-temperature body into low-temperature one without any other changes. For this reason, it is assumed that there are two insulated containers with the same volume and only one wall can conduct heat. The containers contain the same ideal gas with the same number of N molecules. But that temperature T_1 of the gas in the container 1 is higher than that T_2 of gas in the container 2; contact between walls of the two containers after the heat conduction process occurs, both the final temperatures of the two containers becomes T . Both containers have undergone an isovolumetric process, so the Clausius entropy can be expressed as

$$\Delta S_C = \int \frac{dQ_1}{T_1} + \int \frac{dQ_2}{T_2} = \int_{T_1}^{T} \frac{C_V dT}{T} + \int_{T_2}^{T} \frac{C_V dT}{T} = C_V \ln \frac{T_1}{T_2} > 0 \quad (1.25)$$

Clausius entropy is greater than zero, which is an irreversible process.

When discussing this heat transfer process at the macro level with $S_{B,\,Mac}$, it can be seen that the number of macro states of the two containers at different uniform temperatures is $1+1=2$, and the number of macro state of the two containers after heat transfer becomes to 1, in this process $\Delta S_{B,\,Mac} = 1-2 < 0$ and it is completely different and has opposite signs with ΔS_C

If the above-mentioned process is discussed with S_B at the microscopic molecular motion level, it can be seen that although the two containers with temperatures of T_1 and T_2 have the same number N of molecules, they have different energy level combinations and different partition number of molecules at each energy level. After heat exchange between the two containers, occurs, the total energy $E_1 + E_2$ is to remain unchanged. It is equivalent to rearrangement of momentum space configuration from the respective original molecular energy level and molecular partition state on each energy level in $2V$ spaces. Therefore, it can be expressed as

$$\Delta S_B = S_B(T) - S_B(T_1) + S_B(T) - S_B(T_2) = \quad (1.26)$$

$$= 2S_B(T) - (S_B(T_1) + S_B(T_2))$$

When the $2N$ molecules in the $2V$ space reach the thermodynamic equilibrium T and p again, the identical energy level combination and the molecular partition combination of each energy level, that is, the same momentum space configuration, are constructed in the two containers. The two containers have the same number of microscopic states. That is $W[2V(T)] = W[V(T)]$. It is equivalent to a set of spaces with the same distance δV_i in the two containers, which are packed up in the dV space. After the heat exchange, the collision frequency in the two containers has changed, which can be seen as the number of N molecules in the container $V \to 2V$. The total volume has not changed, and the local average number density can be assumed to be $\frac{n_{i,1}(T) + n_{i,2}(T)}{2}$. The total number of particles in2 V space is still $2N$. The total number of microscopic states is determined by the energy levels numbers $I(T)$ determined by temperature T, numbers $I(T_1) + I(T_2)$ of localized locator before heat conduction and the numbers $I(T)$ of localized locator after heat conduction, and then $I(T_1) + I(T_2) > I(T)$ should be certain. In Maxwell distribution of classical mechanics, for the reconstructed energy levels, $I(T) = [I(T_1) + I(T_2)]/2$ can be assumed. It should also be plausible for quantum distribution, since the temperature difference corresponds only to the numbers of molecules n_i in each energy level, without affecting the numbers $I(T)$ of localized locator, i.e., the number of energy levels I or locator domain numbers after conductive, is the same as the numbers of energy levels in one of two containers. Namely the heat conduction process changes the numbers $I(T)$ of localized locator. While the number of particles in each energy level, $n_i(T) = [n_i(T_1) + n_i(T_2)]/2$ for the same energy level, the total number of particles can still be invariant. That is, the microscopic states of the two containers can be calculated according to the energy levels of the one container after the heat conduction process. Or the number of microscopic states produced for molecules of $n_1 + n_2$ in $2V$ space through spatial degeneracy is equal to half of the number of microscopic states produced after heat conduction through spatial degeneracy for molecules of $n_1 + n_2$ in $2V$ space. That is, the number of spatial degeneracy states at each energy level has doubled, but the locator number is still equal to $I = \left(I(T_1) + I(T_2)\right)/2$. In two containers before reconstruction, $W_0 = W_1(T_1) \cdot W_2(T_2)$, the Boltzmann entropy shall be expressed as

$$
\begin{aligned}
S_{B0} &= S_{B,1}(V_{1,0}) + S_{B,2}(V_{2,0}) = k(\ln W_{1,0} + \ln W_{2,0}) \\
&= k \ \ln \prod e^{n_{i1}} \prod e^{n_{ii2}} = k(\sum n_{i,1}(V, N_1, p_1, T_1) + \sum n_{i,2}(V, N_2, p_2, T_2)) \\
&= k\left(N_1(p_1, T_1) + N_2(p_2, T_2)\right)
\end{aligned}
\tag{1.27}
$$

The number of microscopic states reconstructed is as follows $W(T) = W_1(T) = W_2(T)$

$$
\begin{aligned}
& S_B(2N, 2\not{V}, p, T) = S_B(N, \not{V}, p, T) = k \ln \prod_{I(T) = \left(I(T_1) + I(T_2)\right)/2} \\
& e^{n_{i,1} + n_{i,2}} = k \frac{N_1(p, T) + N_2(p, T)}{2}
\end{aligned}
\tag{1.28}
$$

Therefore, the entropy change of the heat exchange process expressed by Boltzmann entropy becomes

$$\begin{aligned}\Delta S_B &= S_B(N,\not{V},p,T) - S_0 = k\ln\frac{W(T)}{W(T_1)\,W(T_2)} = \\ &= k\frac{N_1(p,T)+N_2(p,T)}{2} - k[N_1(p_1,T_1)+N_2(p_2,T_2)] \\ &= -k\frac{N_1(p_1,T_1)+N_2(p_2,T_2)}{2} < 0\end{aligned} \tag{1.29}$$

So far, we have not only obtained two completely different entropy changes for the same natural process, but also with S_B explained that these are two completely different entropies at macro and micro levels. Moreover, two completely different process criteria are obtained. Clausius entropy increase is an irreversible criterion for cyclic processes, while Boltzmann entropy decrease is a criterion for spontaneous processes. That is to say, it is a mathematical expression of Clausius' statement of the second law of thermodynamics.

$$\Delta S_B < 0 \tag{1.30}$$

It holds true for the spontaneous direction of all natur processes. All the problems involved in this book also want to prove its universality.

From the above analysis, it can be concluded that the heat conduction process is a completely deterministic energy transfer process at the macro level, and the distribution of molecules at various energy levels before and after is uniform and orderly in the presence or absence of wall space at the micro level. The only difference is that the number of energy levels has changed and molecules of different energies of each container have been recombined in the space with $2V$. In this rearrangement process, the momentum exchange of particles at the microscopic level at the same spatial position completes the molecular number partition at the temperature of T. The microscopic interactions in the two containers are also asymmetric and random. Therefore, the randomness of micro-level interaction is the source of irreversibility. The above results will also be obtained when the classical irreversible processes of mixing and frictional heat generation are discussed below. These processes are all related to irreversibility, but they are not necessarily spontaneous processes.

The process in reality is a “one-way car” in nature, especially life. In the real world, the criterion of cyclic process cannot be applied. Even if a reversible cyclic process can be designed, it is still impossible to judge what process is spontaneous. This is also why Gibbs proposed the spontaneous criterion of minimizing free energy. The discussion of the mixing process in the next section is a classic case.

1.6. Gibbs Paradox and Mixed Entropy

The mixing process is the same classical spontaneous process as the heat conduction from high temperature body to low temperature one. These processes can be said to be everywhere in nature. Nowadays, you can see very few pure substances like gold and diamonds. Not to mention that the biological genes are constantly being fusion. Therefore, discussing of this problem is also a proof of the statement about the spontaneous direction of the thermodynamic second law.

Different from the thermal effect of chemical reaction heat, the mixing of two ideal gases has mixed thermal effect that is also considered without any interaction. The Gibbs paradox caused by such entropy has a more special generality.

The difference of Clausius entropy results of mixing between the same ideal gas and between two kinds of different ideal gas having the same molecular number at constant temperature. Gibbs proposed the famous Gibbs paradox. He introduced the giant canonical statistical ensemble to explain Gibbs paradox. This ensemble is a microelement at the level of continuous medium. When the ensemble number is large enough, it is still a statistical system. The giant canonical ensembles permit heat and mass exchange between ensembles each other. Such a system can include chemical reactions and phase transitions, but it can avoid explaining the difference between ideal gases and liquids and solids. Boltzmann's statistical method based on molecular motion can still be used.

The ensemble is regarded as a statistical unit, which extends the concept of Boltzmann entropy. It is very valuable to extend the concept of free energy to condensed state. In the following discussions, we can see that this expansion is well-founded. Unfortunately, there is no distinction for divisibility between momentum and space. Still, six variables are combined into one phase space.

The concept of "ensemble" is indeed valuable, but this is not necessary. The mixing problem can still be discussed at the molecular level.

1.6.1. Clausius Entropy Change and Boltzmann Entropy Change in the Mixing Process

Two containers with the same volume are respectively filled up by ideal gases A and B. The temperature and pressure are the same in two containers. After the middle spacer plate is removed, the mixing of molecules A and B can be regarded as an expansion under isothermal and isobaric conditions. According to the reversible expansion process,

$$pdV = \frac{nRT}{V}dV = TdS_C \tag{1.31}$$

The Clausius entropy of molecules A and B is respectively

$$\begin{aligned} \Delta S_A &= n_A R \ln \frac{2V}{V} \\ \Delta S_B &= n_B R \ln \frac{2V}{V} \\ \Delta S_{mix} &= (n_a + n_B) R \ln 2 > 0 \end{aligned} \tag{1.32}$$

This is the entropy change calculated according to the heat of reversible expansion process. The Clausius entropy of the mixing process increases, so it is irreversible process

If Boltzmann entropy is used to discuss the above-mentioned mixing process, it is the process that the molecular momentum space configurations of A and B are transformed from V to $2V$ space. This is a process in which the thermodynamic system itself changes. Boltzmann entropy changes under isothermal conditions. Under the condition of keeping the total energy unchanged, the original particle number combination at each energy level still relies on the momentum exchange generated by complete elastic collision when changing into $2V$ space. Finally, the momentum space configuration reconstructed by molecules A and B in $2V$ space is still uniform. The number of molecules of each energy level shall be $\left(n_{i,A}+n_{i,B}/2\right)$ in each space position and each mixed microscopic state of molecules A and B, only in this way can the interaction p produced by collisions with the same number density remain unchanged. Different δV_i are packed together in each space position, keeping the temperature T at each position uniform and constant. Therefore, the number I of molecular energy levels after mixing is the same as the number of energy levels I of A and B before mixing. Therefore, when calculating the total number of microscopic states after mixing, only the sum on I energy level is obtained. The mixing process changes the numbers of locator in the overall system. The locator number for the mix alters changes from $2I$ to I . But the number of particles at the same energy level in the same locator may be unequal due to large differences in quality, i.e., $n_a \neq n_b$. In this case, the distance space formed by the two particles may not be a regular tetrahedron, and the δV_i after the mixing process may be changed. It is also possible that when the volume is not changeable, areas with pressure difference occur. Such as occurs in osmotic pressure. The number of particles in the spatial configuration δV_i of this energy level in the local locator after mixing should therefore be $\left(n_a+n_b\right)/2$ at the simplest case under conditions of invariant pressure and total volume. And the number g_i of spatial degeneracy states for each same energy level in $2V$ space is also equal to $\left(n_a + n_b\right)/2$. When the number of locators decreases to I , the number of microscopic states in the whole space is $W_{A+B} = \prod_{I} e^{(n_{i£,a}+n_{,\,,B})/2}$. Then the statistics about microscopic states for $2I$ local locators with the same energy level before mixing can be performed. Thus, the microscopic states before and after the above-mentioned mixing are written as the formulas (1.33) and (1.34)

$$
\begin{aligned}
S_{B0} &= S_{B,A_0}(V_{A,0}) + S_{B,B_0}(V_{B,0}) = k_B(\ln W_{A,0} + \ln W_{B,0}) \\
&= k_B \ln \prod_{I} e^{n_{ii,A}} \prod_{I} e^{n_{i,B}} = k(\sum_{1}^{I} n_{i,A}\left(V, N_A, T, p\right) + \sum_{1}^{I} n_{i,B}\left(V, N_B, T, p\right)) \\
&= k_B\left(N_A\left(V, p, T\right) + N_B\left(V, p, T\right)\right)
\end{aligned}
\tag{1.33}
$$

$$
\begin{aligned}
S_{A+B} &= k \ln W_{A+B} = k \ln \prod_{I} e^{\frac{n_{i,A}+n_{i,B}}{2}} = k \sum_{1}^{I} \frac{n_{i,A}\left(2V, T, p\right) + n_{i,B}\left(2V, T, p\right)}{2} = \\
&= k \frac{N_A\left(2V, T, p\right) + N_B\left(2V, T, p\right)}{2}
\end{aligned}
\tag{1.34}
$$

$$
\begin{aligned}
\Delta S_{B,mix} &= S_{B,A+B}(\left(N_A + N_B\right)/2), V, p, T) - [S_{B,A_0}\left(V_{A,0}\right) + S_{B,B_0}\left(V_{B,0}\right)] \\
&= k_B \ln \frac{W_{A+B}(2V)}{W_A(V)\, W_B(V)} = -\frac{1}{2} k_B\left(N_A + N_B\right) < 0
\end{aligned}
\tag{1.35}
$$

Mixing is a classic spontaneous process, and there are also

$$
\Delta S_{B,mix} < 0 \tag{1.36}
$$

On the molecular motion level, the spontaneity of the mixing process of the two matters, which is very certain on the macro level, can be clearly illustrated.

Because the number of energy levels I is unchanged before and after mixing, the spatial degeneracy state number $g_i = \left(n_{i,a} + n_{i,a}\right)/2 = n_{i,a}$ and molecule number $n_{i,a} = \left(n_{i,a} + n_{i,a}\right)/2 = n_{i,a}$ in each locator will not have variety for the mixing of two identical particles. The geometric spatial configuration of distance space δV_i does not change. Therefore, there isn't mixing entropy change for two same kinds of particles. The Gibbs paradox can only reasonably be interpreted with the concept of spatial degeneracy states.

At the same time, the randomness accompanied by determinacy in the mixing process is the source of the irreversible process. Mass diffusion process on continuous medium level is described according to respective concentration gradient. Even the three-dimensional diffusion of space is a deterministic material flow, until there is no concentration gradient everywhere. However, at the level of molecular motion, particles shall not only be uniformly distributed everywhere, but also the number of particles in each energy level of the molecule shall be uniformly distributed everywhere in the space with wall. And that what is not needed when the same molecular mixing. It just is the root of Gibbs paradox. This process of particle energy redistribution is a random process completed by the random motion of molecular collision.

This is to explain once again that irreversibility has no causality with the spontaneity of natural processes. They are two different concepts. A correct interpretation of the mixing process is of great value for understanding many natural processes, especially the generation and evolution of life.

1.7. Statistical Problems of Thermodynamic Systems

Thermodynamics is essentially a statistical problem, whether classical or quantum. No matter the F-D or B-E statistics of single particle or macro system M-B statistics, they are all in order to give determinacy. However, the statistical problems of thermodynamics have not been completely solved. But it does develop in doubt.

Maxwell-Boltzmann statistics is the statistics of obtaining the particles partition according to the energy level for the momentum of molecules, which is based on the divisibility of momentum. But on the premise of space uniformity, the momentum of the gas molecule is an off-domain which cannot be distinguished in the space; there is a locator statistic only for solids. The problem of Maxwell's demon is a query on the assumption of space uniformity. And what really raises this question is the Gibbs paradox of gas mixing. The core problem is the spatial discernibility of momentum in statistics. The spatial divisibility before mixing becomes indiscernible for space. So Gibbs proposed ensemble statistics. This is a statistical method with both momentum and spatial divisibility. But it is described in the form of phase space on the level of continuous medium. The average for phase space smears out the spatial divisibility. But why is it not discussed at the molecular level? This involves the infinite division of continuous coordinates in absolute space. The potential well of the quantum state of momentum in quantum mechanics solves this difficulty. According to the solution of Schrodinger equation, the quantum state depends on the size of space, which is consistent with the concept of "distance space". However, the "space" of Schrodinger equation is still an absolute space in essence and has nothing to do with motion itself. Statistical problems have been wandering between absolute and distance spaces.

Any force can construct the distance space between particles. For the same identical particles, according to distance space, particles with different energy levels are locator sub-statistics of distinguishable particles as a whole space. They are local locators for different distance spaces, the number of the locator is equal to the energy level I , and these local locators are irrelevant to whether the locators are uniformly distributed in the space or concentrated at each local space in the whole space. This is in contrast to absolute space, which results in a new local off-domain sub-statistical approach.

In the F-D statistics of single particles, the space resolution (such as electron orbits in an atom) is determined by the number of spin quanta, but only still consider the microscopic states of all particles on all the space orbits are considered in the statistics. The orbits i.e., a kind of distance space, are virtually ignored. The distance in the electron cloud is a random statistic amount. The statistical certainty is the corresponding different orbital radius values in the spectrum. This is self-consistent with the solution of Schrodinger's equation. Single particle quantum statistics also require the concept of distance space. The specification of the spin quantum number should be viewed as a limitation of the number of particles in distance space. This just is the difference between absolute space and distance space in statistics. Only distance space has the concept of spatial degeneracy states.

The concept of "spatial degeneracy" is equivalent to the concept of "commutative degeneracy" for position in F-D statistics. The concept of degeneracy unlimited particle number for energy band in the B-E statistics is also included. Therefore, the concept of "spatial degeneracy" in local off-domain sub-statistics is the most important core that three statistical methods of thermodynamics can be unified.

The particles with different momentum have divisibility, and the space has also divisibility; both are deterministic primary causes of thermodynamic systems. It is crucial to find this certainty with the concept of "spatial degeneracy states". Thermodynamic systems have only randomness in local space.

Therefore, the importance of understanding the concept of "distance space" from a statistical perspective lies in its universality to thermodynamic system. This concept can be applied at the level of molecular motion, and also at the level of continuous media. It is also applicable to subatomic levels, such as the electron orbital space of atoms.

With the help of the certainty of motion at a certain level, the uncertainty of micro level statistics motion at a smaller scale is a reasonable for irreversibility.

Another problem closely related to the concept of distance space is the time of ergodic processes to reach the thermodynamic equilibrium state. There are thermodynamic equilibrium parameters everywhere in the non-uniform high-speed flow field of aerodynamics. It can also maintain a lasting stable equilibrium. In other words, fluctuation does not matter. Therefore, it must be said that thermodynamic statistics shall not have the time of ergodicity. The ergodic time has nothing to do with the motion itself. The time of contact with motion refers to the time when the particles subjected to the action, such as after the two particles collide, the momentum is changed to reach a new distance. This can't be ergodic process. The ergodicity violates the thermodynamic second law, $\Delta S_B < 0$ for natural processes, that is, the number of states experienced in time series should also be minimized. When the temperature changes, TS_B is showing that this potential energy has the ability to complete the reconfiguration of energy allocation in a minimum time. Reconstructing a distance space with collision, this time is the time during the pressure changes, and the pressure response is the response of interaction to reconfiguring the distance space. The Poisson equation of pressure indicates the global response of pressure field and the instantaneity of pressure response.

1.7.1. Statistics of Systems

1. Systems that only give the number of particles and total energy of the system are uncertain, because the distance space of the system without the specific force is uncertain. 2. The statistics of a thermodynamic system are then derived from the uncertainty of δV_i and the corresponding population dn_i and spatial distribution. The certainty arises from thermodynamic equilibrium conditions (thermodynamic parameters p and T are

homogeneous everywhere). But the introduction of the concept of "spatial degeneracy states" in local off-domain helps us to find such certainty.

The variety of the world originates from the statistics of the thermodynamic system, and the real things are the realistic mapping of various thermodynamic system. From a statistical point of view, we also need to abandon the traditional expression of absolute space time.

Whether it is microscopic molecular motion or macroscopic phenomenological thermodynamics, the system without potential energy is incomplete. The thermodynamic system without TS_B potential energy is a non-consistent system.

1.8. Adiabatic and Isentropic

In the discussion of adiabatic heat conduction process in section 1.5, two entropies are obtained. Although it is an adiabatic process for two containers, however, this is not generally considered to be an adiabatic process in the usual sense. The expansion or compression of a gas system under adiabatic conditions is called adiabatic process. Under reversible conditions, this adiabatic process is called the only-isentropic process.

Imagine two insulated containers with insulated piston and both have initial volumes of V_0, they were both filled with ideal gas at the initial temperature of T_0. The gas pressure in container 1 is p_1 and in container 2 is p_2. Connect piston each other with a rod in the middle. Under the action of pressure difference, the pressures of the two containers reach the end pressure p. The volume of the two containers becomes volume V_1, and V_2 respectively.

It can be seen from the initial state that the number of molecules in the two containers is not equal, have unequal relationship $S_{B,1} \neq S_{B,2}$. Therefore, the adiabatic expansion and adiabatic compression lines on the $p-V$ figure are two different lines. So that the p_1 expansion line and p_2

compression line on the T_0 isotherm line will intersect isobaric line p to achieve pressure balance that two points corresponding temperature T_1 and T_2 . But there are still two macroscopic states with different temperatures. Even if it is irreversible, that does not expand and compress with infinitesimal pressure difference respectively, because there is no heat exchange, there is still the following formula,

$$\Delta S_C = 0 \tag{1.37}$$

However, if the state of the two containers is regarded as a macroscopic state, then initially there are two macroscopic states for the two containers with different pressures. After adiabatic expansion and compression, they become two macroscopic states with different temperatures. Therefore, there are

$$\Delta S_{B,\ Mac,adi} = 2 - 2 = 0 \tag{1.38}$$

This is an isentropic process.

So, if this is the $\Delta S_B = 0$ case in the spontaneous processes? Considering that there is $T_2 > T_1$ after the adiabatic process, to reach the thermodynamic equilibrium completely, the adiabatic heat conduction process, $\Delta S_B < 0$, is also required. From the end to the thermodynamic equilibrium it is more appropriate to exclude the equal sign from the criterion of spontaneous process. However, please remember that there is also a kind of spontaneous process due to imbalance of pressure or interaction in the system. They also have the tendency to balance, which is also important spontaneous process.

Naturally, the Boltzmann entropy change is also found in each of the two containers. Although the number of particles in these two containers is not equal, the number of microscopic states is not equal, so this problem can still be attributed to the spontaneous reduction of Boltzmann entropy of

$$S_{B,0,1} > S_{B,0,2} \tag{1.39}$$

However, the number of molecules in these two containers has not changed, but the $p{V\!\!\!-}$ potential energy of each system respectively changes its kinetic energy T or the total energy to determine this system in the adiabatic process. Therefore, the Boltzmann entropy of each system has also changed. There should be

$$\left|\Delta S_{B,1}\right| = \Delta S_{B,2} \neq 0 \tag{1.40}$$

Then there is formula (1.38). This example also helps to explain why there is such a correlation for ${V\!\!\!-}$, and S_B that can be written as $S_B\left({V\!\!\!-}\right)$.

Chapter 2

QUESTIONS OF MOTION ON MOLECULAR LEVEL IN CONDENSED MATTER

Man and everything on the earth around him are made of gas, liquid and solid matter. They're all thermodynamic systems constitute of a lot of molecules. Thermodynamic relationships and state functions were originally defined by gases and then extended to all condensed matter. And in doing so, we use free energy, which implies all the thermal effects, including the phase transition heat. But our understanding of phase transitions is so poor; beyond the knowledge of phase laws and phase diagrams, we cannot even answer why there is a large latent heat of phase transition. Knowledge of latent heat is not more than a statement of a "fact" or of common sense.

A variety of critical point experiments, including Bose-Einstein condensates, super-currents, and superconductivity experiments, have gained the Nobel Prize and promoted the study of "critical points" in condensed matter physics. Theory, including average field theory, scale law, universal law and renormalization group, has made continuous progress in phase transition research. But the "phase transition" problem remains still challenging and full of opportunities because there is a lot of experimental "bonanza". But it is not clear whether the theory of continuous phase transitions based on critical points can be applied to discontinuous phase transitions. The "critical point" of the transition of gas and liquid phase is just on the "boundary" between discontinuous phase transition problem and continuous phase transition one. It could therefore serve as a starting point for discussion.

Condensed matter has motion at the level of continuous medium, but phase transition is only concerned with molecular motion in condensed matter physics.

2.1. Phase Change

It is the theme of this chapter to propose some new thoughts on the theory of phase transition with this universal symmetry relation $pV = TS_B$. And we can see from the discussion that phase transitions are the best way to understand intermolecular interactions, including ideal gases. The difference of condensed matter is the difference of the intermolecular interaction. The different types of phase transitions correspond to the mutations in the main momentum types of molecules in condensed matter. The phase change heat is the potential energy conversion amount $\Delta(TS_B)$ during transition of intermolecular interaction.

2.1.1. Review of Critical Point Phenomena and Theories

The first critical point problem started from the critical point of no liquefaction in the phase transition process between gas and liquid, which gave birth to van der Waals equation. The Weiss' theory of "molecular field" was developed by studying the Curie point of the disappearance of magnetism. In place of the interaction with a molecule field of parallel to the spontaneous magnetization that is equivalent to the occurrence of a disordered molecular field in the solid magnet that changes order. Knell found that when the strength of the external magnetic field is large enough, small sub-lattices will suddenly appear in the whole non-magnetic sample during the anti-ferromagnetic phase transition, which is of microscopic magnetic order. On the contrary, X-ray diffraction experiment shows that when a critical temperature of 724K is reached, the lattice atoms of Cu and Zn in the brass alloy are mismatched in their respective cubic lattice nested structures. Both are phase transitions between "ordered" and "disordered" positional spaces. Under the condition of increasing pressure, liquid helium can becomes solid state at very low temperatures. And "super-current" is a kind of behavior of macroscopic translational momentum after forming clusters with relevant scales under special interaction. Super-conductivity is that the translation and rotation momentum of the two charges are zero in the Bose-Einstein condensate state, but there still is an elastic force in the lattice that produces vibration momentum so that the electrons with opposite momentum and charges can be become Cooper pairs. And the force in the lattice causes these Cooper pairs to form macroscopic or mesoscopic clusters showing superconducting phenomena in the crystal. In all these experiments there is a universal feature of a phase change that is caused by a sudden change in force. There are three common features for all the above critical point phase transitions: 1. The emergence of microscopic clusters associated with the concept of related scale 2. There are long-range interactions in the formation of clusters. 3. The change of the cluster momentum type occurs at the force mutation near the critical point.

The concept of "order parameter" is used in various theories of critical point phase transitions. The physical meaning behind this concept is the mutability of critical point phenomena and the great changes in macroscopic systems caused by tiny differences. In other words, the order parameter is mathematically small parameter, but it causes a change in the macro function. This is also the basis for using order parameters in Landau's "average field" theory to construct the solution of the kinetic equation of phase transition. The undetermined function in the construction solution can be determined by fitting the experimental results. This approach is important in that the phenomenological order parameters can't always accurately reflect the "correlation scale" based on molecular motion of the level.

Therefore, some phase change statistical model, include Lenz-Ising model and Heisenberg model, on the micro-particle interaction are presented. By calculating partition function for the possible energy level of the particles in the position space generated by the specific interaction in the phase change problem, a statistical method is used to solve the macroscopic physical quantity of a mutation in the phase transition of the critical point. However, it is difficult to determine all possible combinations about long-range relation and how to truncate the physical conditions of the correlation combinations at what scale is difficult. And, as we've already discussed, what determines the partition of the energy level is also the force that determines the spatial position distribution of the particles. However, it may not conform to the physical reality to allocate energy in the space configuration of the artificial geometric position. Tell colloquially: this is "shoe can fit foot" problem. Choose the higher order series expansion, and then the coefficients obtained by re-calculation are also compared to the experimental or the known coefficient. Without the certainty of macroscopic thermodynamic parameters, it is impossible to have the certainty of statistical physics. This can explain why in the flat ice entropy problem, only the statistics in the real spatial position have accurate solution. And certainty also comes from the fact that the momentum space configuration is not random and cannot be calculated by using the statistical model of random walk. It is assumed that what need of these statistical models is to determine the distribution of momentum in space during phase transition. Make it meet the macroscopic thermodynamic equilibrium and the phase equilibrium conditions on the determined value. And the spatial degeneracy states could greatly reduce the computation. This may make it possible to compute three-dimensional Issin or Heisenberg models.

The strict solution of the two-dimensional Issin model reveals the problems of the mean field theory, namely the logarithmic singularity of the specific heat at the critical point instead of the finite jump, and the $\beta = 1/8$ instead of the 1/2 obtained from the mean field theory. Experimental facts also encounter some questions how the average field theory find a new general statement in place of the original constructive solution for a large number of different critical point phase transition. This raises a question, that is, what should be the good functional form of the solution? How can its variables be determined universally?

It is not only an important method of experimental design by used of dimensionless variable group, but also it is a good way to find universal law from experimental data. The scaling law of phase transition problem does obtain the same critical exponent in completely different critical point phase transitions. These exponents reflect the relationship between the correlation scales produced by interaction varying in the critical phase transition and the characteristic scales of the change of momentum spatial configuration in the phase transition. Because it is an approach method based on experiment, then it is possible to meet the problem above mentioned about matching the shoe to the foot. And these scaling laws are consistent across all critical point phase transitions. This universality is undoubtedly that we hope to look for invariance in the changes for all problems. Experimental research is so powerful with dimensionless analysis alone. The theoretical conclusion is that there is a long range correlation scale δ in critical point phase transition, and the critical temperature scale β can only be related to the dimensions of momentum space, even with the dimension D of space. This universal classification is based on the number of spatial dimensions D and order parameters n, namely n is the number of freedom degrees determined by the number of momentum types, for example translation, rotation, vibration and the spin momentum. The "invariant" found here is that a self-similar scale l can be used as a ruler to measure the scale of clusters by used of l^D in the D dimensional space. This is the new unit of counting the momentum of microscopic states involved in phase transition, namely the number of clusters instead of the number N of molecules. In the physical sense, it makes no sense for a larger scale than the spatial correlation scale if it is not a unit carrier a kind of new real momentum. And the smaller scales than this scale do not become the individual producing the actual momentum exchange. Because the scale set smaller than correlation scales are nested in clusters having correlation scale with self-similar structures. This is the problem of the change about distance space elements δV_i during a phase transition discussed in section 1.4. This result also proves from the point of view of phase transition that ideal gas molecules with different momentums in wall box space to pack up in the same space position according to different scales δV_i are correct. Only this spatial distribution of momentum can explain the spatial "uniform" configuration of gas momentum. Random walk is an incorrect physical model. The randomness of space and time series in the process of phase transition is contrary to the minimization of microscopic states in the $\Delta S_B < 0$ process.

This self-similar structure sets to be packed up is also the physical basis of the singular saddle points with invariance for renormalization group method. It is with this invariant correlation scale that the interaction of smaller correlation scales does not need to be taken into account in the statistical model. The renormalization transformation group is constructed by using the intermediate transformation, that is, these Hamiltonian equations with different characteristic scales are constructed, each of which has the corresponding points in the "order parameter space". The saddle point obtained by calculating the moving trace of point in the ordered parameter space corresponds to the correlation scale related to the critical point in self-similar scale sequences.

There may be less than one fixed point in the ordered parameter space that is related to the degrees of freedom n, namely the number of momentum types, and the spatial dimension D. It is also demonstrated that: the general class types of the phase change are divided according to two spatial configurations.

But is it necessary to compute the relevant scale to infinity in the renormalization transformation? According to the physical meaning of the self-similar scale, the critical point of the corresponding fixed point means that the number of clusters is determined by the actual working correlation scale and larger than the correlation scale is not meaningful owing they are man-made scales. And the short-range forces reach the limit and no longer working, which means a reversal of the interaction. That is, the long-range forces that occur on a larger cluster scales are the interactions in a smaller distance space $\delta V_{i,\,Cluster}$ between clusters particles.

The change of force in phase transition is the most fundamental which changes the number of momentum types, that is, the number of freedom degree n. Based on the n value when the acting force is reversed, the difference between the Hausdorff fractional dimension corresponding to the cluster and the geometric space dimension D is used as a small parameter for introducing a series expansion and constructing a series solution to generate another solution method of a critical index. The result is completely matched with the critical index value obtained by renormalization group transformation. The experimental value of the supercritical phase transition of the liquid helium is also consistent. The consistency of the two theoretical algorithms shows that the clusters contain every related scales and distance space scales, and are nested and filled in the clusters. At a fixed critical temperature, it is the same matter of having a fractal space dimension and renormalization group transformations for the clusters.

In order to better extend the results of critical point phase transition to the discontinuous phase transition between the gas, liquid and solid, it is the most suitable way to start from the critical point of gas -liquid phase transition on the isotherms. The core problem should be how the distance space is such that the force is abrupt.

2.1.2. Phenomenon in the Gas-Liquid Critical Point Phase Transition

In 1869, the British physicist Andrews discovered the critical point of the gas-liquid phase change in the study of gas-liquid phase transition of carbon dioxide gas.

Experimental observation was carried out with three floating balls with densities less than or equal to and greater than the critical density of carbon dioxide gas in a glass tube sealed some pure carbon dioxide gas. In the process of cooling, it can be observed that when the temperature is closed to the critical point and the critical pressure is maintained, three spheres of different densities are separated and located in the top, middle and bottom of the glass tube. This phenomenon can occur when the gas density is equal everywhere, everywhere equals the critical point density, or it may occur under the condition that the gas density is different everywhere, and the floating balls with different density stop at the positions in gas phase with same density. When the temperature drops to near the critical temperature, the "critical opalescence" generated by the enhanced light scattering can be observed. When the temperature drops, the density of the gas becomes usually large, but the floating ball in the middle is sinking, and the gas-phase density shows to become uneven from uniform or not more uniform from the non-uniform. Once the temperature decreases to the critical temperature T_C, the phase interface appears immediately, the float ball with smallest density at the top falls to the liquid surface and the gas phase density decreases compared with before. The float ball equal to the critical density rises to the phase interface. As the temperature continues to drop, the critical opalescence disappears in the gas and liquid two phases and the bottom float ball rises to the liquid level with the increase of liquid phase density, and all three spheres float on the phase interface.

Critical opalescence is caused by the scattering of light on particles that are close to or larger than the light wavelength. Light does not scatter on particles at the molecular scale, so conventional gases and liquids are transparent to visible light. It shows that the cluster scale near the critical point is comparative to the wavelength of light. Critical opalescence is the most direct evidence of clusters in gas or liquid. Whether the gas density is or is not uniform before the approaching the critical temperature, after the appearance of the cluster, the sinking of the floating ball equal to gas density of the critical point indicates that the gas phase density above the cluster in the gas phase becomes smaller and the gas phase density above it is no longer uniform that can be determined. The initial position of the three balls can also be caused by the uneven spatial distribution of gas density due to formation of cluster near the critical temperature.

Clusters and correlation scales are almost the focus of all theories of critical point continuous phase transition. But why do molecules without polarity also have clusters? What is the relationship between discontinuous phase transitions and clusters? None of this is clear.

Using the symmetry of $p\not{V} = TS_B$ to re-examine the existing these phase transition experiment, the conclusion can be summed up as follows: 1. The essence of phase transition is the change of the momentum type of dominant kinctic energy in condensed matter, i.e., the change of freedom degree of; 2. According to the above symmetry, The change of condensed matter is the result of the change of some interaction force. Without potential energy, there's no way to talk about the force, no way to talk about the phase transition. Where there is interaction, there is no "permanent gas" and "permanent liquid"; 3. The appearance of opalescence indicates that cluster formation is a precursor to phase transformation. And the appearance of certain clusters is a precursor to phase transition, a precursor to the reversal or mutation of the interaction force. Correlation scale is related to cluster scale 4. The cluster is related to how many levels of distance space δV_i are packed up in them.

2.1.3. Momentum of Rigid Diatomic Molecules in a Thermodynamic System

For rigid diatomic molecules, there are three kinds of momentum. Classical mechanics is used to describe the momentum in geometric space in which a molecule has a momentum of translational motion with the total mass $m_a + m_b$, and rotating momentum as a rotator with a reduced mass $\mu = m_a m / (m_a + m_b)$ rotating motion around the molecule mass center with the moment of inertial $I = \mu r^2$ and the vibration momentum of the resonant vibration motion on distance r between the diatomic nuclei. The molecular momentum is given by microscopic quantum momentum, then the quantum state description of the degree of freedom composed of translational quantum number n_x, n_y, n_z, rotational quantum number j, magnetic quantum number m and vibrating quantum number v_x, v_y, v_z

$$\langle n \rangle = n_x, n_y, n_z \; j, m, v_x, v_y, v_z \tag{2.1}$$

In the phase transition of three states of matter, the electron and electron spin and nuclear spin quantum number in atoms and molecules can't be taken into account. The momentum of these types also has the corresponding distance space element δV_i and the corresponding two kinds of potential energy.

2.1.4. Gas-Liquid Phase Change

The connection between the gas-liquid phase change and the critical phase change at the critical point is more convenient to illustrate the discontinuous phase change.

The velocity of an ideal gas molecule is the translational momentum of the molecule. The collision of translational momentum in gas produce an interaction p , often referred to as pressure. The macroscopic description of the gas-liquid phase change is to use the isotherm in $p-V$ graph. The isotherm in the figure is a hyperbola above the critical point. The temperature below the point decreases with the pressure drop and the volume expands. When the isobaric horizontal line appears on the isotherm, it indicates the appearance of the gas and liquid phase, which is called gas-liquid "intermittent phase transition".

According to the discussion of distance space unit δV_i in section 1.4, when the scale is larger, it means that the collision of larger molecular momentum will cause the pressure to increase. But pressure in the actual process on the hyperbola is reduced. This is because the heat absorbed in the process increases compensates for the increase in δV during when the pressure drops.

The phase transition is a process of constantly updating thermodynamic system and achieving new balance. There are three changes in the system after the condensation of the liquid phase. 1. The liquid level becomes a movable interface, and the gas phase space is no longer a box space with a fixed wall; 2. In the gas phase, the distance spaces δV_i are no longer uniformly distributed in space. It becomes the distribution with a δV_i scale increase from the height of liquid surface. 3. The gas density, or numerical density, is less than the original gas density.

2.1.5. Cluster and Internal Pressure

The particles near the liquid level have small momentum, high particle number density and high collision frequency. These conditions for forming clusters are also consistent with the opalescence phenomenon at the critical point. Further away from the critical point, opalescence appears in a more thin space, and may not be visible at all macroscopically. This is where the van der Waals equation comes into play. It generates the internal pressure in the van der Waals equation, which is $\left(aN^2/V^2\right)$. The gas molecules momentum with smaller distance spatial scale δV_i on the lower part within the gas phase is relatively small. The pressure p in the liquid refers to the collision force between molecules and clusters, namely the pressure in the clusters during liquefaction. Low momentum lacks the collision of molecules with large momentum, which is similar to the scene when multiple billiard balls hit nine balls at the same time. The formation of this internal pressure is positively correlated with the number density. If these nine spheres can be broken apart, they will still be gas molecules. When the number density in the cluster becomes so large that it can no longer be broken apart, the molecules will gather into clusters, namely the opalescence generated by the clusters during the critical phase transition. Due to the large mass and small speed of the cluster, the cluster still has momentum comparable to the molecular momentum. The internal pressure $\left(aN^2/V^2\right)$ generated by the cluster collision can be the same as that pressure p generated by the upper completely elastic collision of molecule, so as to maintain the condition that the overall interaction force of the system is balanced between phases. In the space dominated by clusters, the internal pressure becomes the interaction force generated by elastic collision between clusters, and eventually replaces the elastic collision force p of ideal gas molecules in this region, namely $p \to 0$. When the translational momentum of the molecule is completely frozen, that is, when the gas and liquid phase change occurs. This $(V - Nb)$ means that there is still distance space $\delta \not{V}'$ between clusters to produce collisions, which will undoubtedly be much smaller than the original scale $\delta \not{V}$. When the number of clusters becomes large enough, that is, when $(V - Nb) \to 0$, the clusters are condensed into liquid state. The molecular translational momentum in this part of the liquid becomes zero.

A change in interaction means a symmetry change. The interaction $\left(aN^2/V^2\right)$ in the liquid phase is the collision between clusters, and the distance spaces $\delta \not{V}_{i,L}$ generated in cluster are the spatial arrangement for transferring the rotational momentum, which also means $\delta \not{V}_{i,L}$ of different sizes corresponding to the different rotational momentum. In the cluster, it is still the molecular collision that carries the transfer of the molecular rotation momentum. The macroscopic statistical average of the interactions of rotation momentum in a liquid resulting from a molecular collision is the shear stress τ_{ij} in the liquid at the continuous medium level. (This problem is discussed in the section on concentrated vortexes in chapter 3). Therefore, the liquid cannot produce large volume change, has the ability to withstand external pressure, and has very small compressibility. The compression ratio of the liquid indicates that there is still some space between molecules in clusters for the transfer of rotational momentum. The pressure p in the liquid at the continuous medium level is the flow of the whole macroscopic translational motion of the liquid. It can also give rise to the rotational motion of the concentrated vortices about a common axis. These forces correspond to different momentums, the pressure p generated by molecular collisions corresponds to the rotational momentum of the molecular level (including clusters), and the shear stress of the continuous media layer corresponds to the macroscopic vorticity.

2.1.6. Phase Interface

The shape of the phase interface reflects the different of interaction force in the two-phase, although it is also influenced by the wall to different two phases. The two-phase pressure is maintained equal by means of the surface tension in the interface layer under phase equilibrium, and is compensated by different shapes. In phase transformation, the phase interface is a constantly renewed surface. The molecules in the transferring process become from the two-dimensional structure $\delta A_{i,L}$ to a three-dimensional structure $\delta V_{i,t}$. This is the cause resulting in the interfacial tension or surface energy.

2.1.7. Phase Change Heat and Phase Change Entropy

Taking gas-liquid phase transfer as an example, ideal gas molecules from the original uniform distribution of microscopic state number becomes to the corresponding α and β phase of the new combination number of microscopic state for two phases respectively, that is the process $W\left(\left(n_\alpha + n_\beta\right)/2\right) \rightarrow W\left(n_\alpha\right) W\left(n_\beta\right)$. The overall condensation phase transition process is the Boltzmann entropy increase process. However, for condensed gas molecules, microscopic state numbers of molecules with small translational momentum are counted according to cluster numbers after the molecules are incorporated into the clusters, so that the gas-phase clusters appears $W\left(n_\beta/2\right) \rightarrow W\left(n_\beta/2b\right)$ and is the Boltzmann entropy reduction process.

$$\left(\Delta S_B\right)_{G \rightarrow g, clus} = k \ln 1/b < 0 \quad b >> 1 \tag{2.2}$$

The abrupt interaction occurs when the gas cluster number density increases sufficiently to freeze translational momentum of the clusters, resulting in the freezing of the N_β/b clusters into liquid phase clusters. The translation kinetic energy T_t of the cluster molecules before being frozen is converted into a rotation momentum T_r of the liquid-phase cluster

$$\begin{aligned} S_{B,cluster}\Delta T &= S_{B,G}\left(T_{G,clus,t} - T_{G,t}\right) + S_{B,L,Cluster}\left(T_{L,r} - T_{G,clus,t}\right) \\ &= T_{G,clus,t}\left(S_{B,G} - S_{B,L,clus}\right) + S_{B,L,clus}T_{L,r} - S_{B,G}T_{G,t} \\ &\approx \left(S_{B,G} - S_{B,L,clu}\right)\left(T_{G,clu,t} - T_{G,t}\right) \end{aligned} \tag{2.3}$$

The first term in the formula (2.3) can be calculated as (2.2) entropy change at the time of gas cluster formation. Although the different types of molecular momentum between two phases but kinetic energy can be assumed to approximate equal. So that only if the molecular kinetic energy $T_{G,t}$ is slightly larger than that of the trapped cluster $T_{G,clu,t}$ in the gas phase, a large phase change heat will be generated.

$$Q_{Conden} \approx \Delta\left(TS_B\right)_{Conden} = \Delta S_{B,G\to g,clus}\Delta T_{G\to g,clu} << 0 \tag{2.4}$$

This may explain how much the condensing phase change latent heat will be so large. This seemingly contradictory assumption just reflects a kinetic factor when the phase change occurs for a thermodynamic equilibrium system with two phases.

Clausius entropy calculated by phase transition heat has

$$\Delta S_{C,Conden} = Q_{Conden}/T < 0 \tag{2.5}$$

However, from the macroscopic view, the condensation changes from single gaseous state to three states of gas, liquid and phase interface in the system, and $\Delta S_{B,Max} = 3 - 1 > 0$. But this does not mean that the reverse evaporation process is a spontaneous process. This is because the phase change system is a non-conservative system in which the total energy is constantly changing. And two opposite discontinuous phase changes are asymmetric, and only one spontaneous symmetric breaking direction exists. This is briefly discussed in the last chapter.

2.1.8. Overcooling and Overheating

The generation of overcooling and overheating phenomena in phase transition is caused by uncertainty of cluster size and random uncertainty of transition between cluster and molecule. The transformed momentum from the molecules to the cluster or a reverse change occurs on both sides of the phase transition process. For example, upon condensation, gas molecules are converted from molecular translational motion to cluster translational motion until the momentum is frozen and converted to rotation momentum of liquid phase clusters. The two corresponding cluster sizes are not determined in the process, and the difference between the two sizes becomes larger as the condensation temperature is reduced, so that the transition between the two sizes becomes difficult, thereby increasing the uncertainty, which is the most main reason for generating the overcooling and overheating. In addition, the randomness of the transition from a two-dimensional structure to a three-dimensional structure also increases as the cluster size increases.

The molecules themselves have momentum distribution, and there are collisions between molecules and between clusters, and collisions between molecules and clusters, so that there is no so-called fluctuation. The uncertainty mainly comes from the uncertainty of the cluster size, and the phase change temperature changes the difference value of cluster size between two phases for vapor-liquid and the cluster scale in the two-phase, so the phase change temperature has an influence on overcooling and overheating.

The degree of overcooling and overheat are affected by external disturbance force stronger than the interfacial tension, such as the intensity of external vibration. The magnitude of the external force relative to the acting force between molecules and the internal pressure influences the overcooling and overheating phenomena, and can also affect phase change.

Phase transitions are transient processes that occur instantaneously, rather than undergoing lengthy ergodic processes. Such as solidification of the solid, too rapid cooling can lead to defects and flaws, even amorphous solids; but none are lengthy. There is ordering of the distribution of momentum in the space of a system, $T\Delta S_{B,G\leftrightarrow L}$ has the ability to quickly restore the system equilibrium in phase transitions. The rate at which the phase change heat is released and absorbed has an effect on the phase change and therefore also on the overcooling and overheating.

2.1.9. Van der Waals Equation

Many of the core concepts in existing phase transition theories are derived from the Van der Waals equation, so it should apply not only to gases but also to liquids and solids. It can help us unscramble the process of liquid-solid phase transition and gas-solid phase transition.

1. $p\not{V}$ The interaction caused by molecular elastic collision has been discussed in ideal gas. It produces the exchange of molecular translation momentum. Other types of molecular momentum in liquids and solids, through elastic collision interactions, all can be considered as p , but does not play a controlling role. It should mainly be reflected in the distance space architecture through the molecular momentum transfer within the cluster. The volume V is the volume of the different distance space of the gas molecule and should be $V\left(\delta\not{V}_g\right)$. In addition to gas, the phase change produces liquids and solids that invariably have $V \neq \sum_l \delta\not{V}_i$, theoretically always leaving an infinitely diluted equilibrium gas phase, but expanding pV into the liquid and solid phases is much more important. This also illustrates the importance of the concept of distance space. Only thereby can Van der Waals equations be applied to both liquid and solid phases. The

δV_i can also extend δA_i to two-dimensional planes or δL_i to one-dimensional line space. Each also has a corresponding spatial position configuration potential.

2. $a(N/V)^2$ Internal pressure is the interaction force between clusters generated by the collision between clusters that has a self-correlation scale l^D. It is also equal to the interaction between the molecules in the cluster. There is still space ($V - Nb$) outside the clusters with greater momentum than the cluster, and the elastic collision between the molecules and cluster is also definable elastic collision p.

The size of cluster is positively related to the number density. The distance space $\delta V(l^D)$ between clusters should also be positively correlated with the number density of molecules. Therefore, the internal pressure generated by the collision of clusters is proportional to the square of the number density. As a result, with the increase of the number of clusters, the ability of the internal pressure to freeze the translation momentum of clusters and molecules between the clusters, and the gas-liquid phase change will appear. Considering that there is no upper limit to the increase in number density, this internal pressure can be applied to the clusters of liquid and solid during phase transition. Liquid and solid all have a concept of internal pressure associated in relation to the respective corresponding momentums.

The internal pressure in the liquid transmits the rotational momentum between the clusters. Although the distance space becomes smaller, it still makes the liquid to have deformability. The thermal expansion and cold contraction of liquid as well as the compression deformation rate are also evidence of this distance space. As the temperature of the step decreases, the distance space becomes smaller, and when the internal pressure freezes the molecular rotational momentum, the liquid phase becomes solid.

Only the vibration momentum of molecules and clusters remains in the solid. There are also clusters in solids, often called “domains”. There are also molecules between clusters, and are also molecules inter-force of each other and force between molecules and clusters. This interaction belongs to molecule Hooke force, which produces molecular interaction p in solids. The interaction between the clusters is that the internal pressure. There is also a distance space $\delta \not{V}_{i,S}$ between the solid clusters. When the internal pressure in the solid is greater than the pressure p , the molecular vibration momentum will also be frozen. The absolute zero-degree temperature may to reach until all other momentums to be frozen.

The internal pressure $a(N/V)^2$ should be the force of freezing a certain momentum on a certain distance space $\delta \not{V}_{i,}$ between clusters in phase transition.

To say whether a certain momentum is frozen or not does not mean that there is no other momentum at same time, but that in this distance space other momentum does not take part in transfer.

The internal pressure in three condensed matter is not in one order of magnitude, but is positively correlated with the square of molecular number density. The internal pressure in the solid is much higher and can withstand very high external pressure.

The internal pressure in the liquid and solid phases is not on the same order of magnitude, and is positively correlated with the square of the number density of molecules. The internal pressure in the solid phase is much higher. That's why solids are subjected to very high external pressure.

The external pressure in the phase transition is the response to Van der Waals forces $[p-(N/V)^2 a]$, and the effect of the internal pressure in the cluster is to reduce the scale of corresponding distance space $\delta \not{V}_{i\,L}$ and $\delta \not{V}_{i,S}$. For example, when you add 20 atmospheres of pressure to "permanent liquid" helium, it turns into solid helium. At a rapid rate of compression, the crystalline ice can be converted to dense amorphous ice. When the distance space $\delta \not{V}_{i,S} \to 0$ it may indicate some interaction to be out of the category of condensed matter physics based on molecular motion.

3. $V-Nb$ According the point of view about cluster, the region where the number of clusters is the largest, i.e., with the largest volume Nb of

molecules. And it is the region having least space of free molecules, i.e., the minimum $(V - Nb)$ region. But there is both a generalized p , which can be defined by the molecular collision, and an internal pressure $a(N/V)^2$. For molecules in gas phase there should be as follow

$$V - Nb = \not{V}_G = \sum_i (\delta \not{V}_G)_i \tag{2.6}$$

This concept can also be extended to liquids and solids, since there are clusters related to the correlation distance space. For example, liquid water has a high density and a low density, and solids of the same material have different crystals and amorphous solids. This can only be changed along different phase transformation paths if the clusters before phase transformation are different, then $(V - Nb)$ and internal pressure is different.

The Van der Waals equation is expanded, so that the liquid-solid phase change and the gas-solid phase change are facilitated to be digested. In the phase change, the TS_B potential energy is necessary to achieve the re-equilibrium between two phases and to reapportion the number of particles at each energy level for relative momentum. Especially explaining the phase equilibrium of different momentum in discontinuous phase transitions is not negligible!

2.1.10. Solid-Liquid Phase Changes

The variation of the force and the scale of the distance space element are explained above. But the effects of external pressure and its loading rate, as well as temperature and temperature change rate on the growth of crystal or amorphous materials, need be specifically discussed.

The external pressure makes the distance space scale in the liquid cluster smaller, and the molecular rotational momentum is more likely to be frozen, and the condensation process is more likely to occur. The effect of the rate of increase in pressures whether the phase change heat can be dissipated in time, that is, whether δV_l caused by the rotational momentum transfer rate in the liquid cluster can be regularly nested assembly, whether the rate of rotational momentum transfer in the liquid cluster can be properly folded, so as to form a regular crystal when the Hooke force (or the Coulomb force between ions) freezes. Too fast cooling and too fast loading will make the production of amorphous materials. Or there are flaws in the spatial structure of the crystal. Therefore, the release of latent heat of phase transition becomes the core problem of crystal growth.

The pressure changes on two sides of the solid-liquid phase interface resulting discontinuously in the distance space scale that generates surface tension. Then the force of two phases can still be kept balance. The amount of molecular with vibration momentum in the solid phase can be redistributed; the temperature of the crystalline phase and the liquid phase are kept equal by the increase in density.

The three-dimensional position configuration of the liquid phase varies to the two-dimensional configuration of the plane of the interface phase, and from the solid two-dimensional interface phase changes to the three-dimensional structure of the solid phase. The momentum space configuration of the surface also creates a potential energy difference. The potential energy change caused by the change of the momentum space configuration is only required to be for the change of the distance space configuration of the regular triangle

$$T\Delta S_{B,\delta V \to \delta A} = kT \ln \frac{6}{5} > 0 \tag{2.7}$$

There is an opposite variation in spatial dimensions, but the two ways will not be symmetrical. The liquid phase and the solid phase both have cluster forming processes, but the cluster sizes of different momentum types are different. The structure within the cluster may or may not be regular, and the degree of regularity between clusters is uncertain. The change of microscopic state occurs together with the change of momentum type, and the variety and their crystal structure are the easiest to detect, so the condensed state physics is an important physical concern field.

The rate of pressurization and rate of temperature change are related to the release of latent heat upon phase change and the rate of heat transfer kinetics. Therefore, the phase change problem is not a pure thermodynamic equilibrium problem, and the heat conduction and the reconstruction of the momentum space are dynamic processes in time evolution. The temperature control technique for crystal growth, such as the technique for obtaining high-quality single crystal silicon, the vapor phase and liquid phase epitaxial growth techniques, is known from long-term practice. These sophisticated temperature control techniques essentially match the rate of the momentum rebalancing by means of the TS_B potential energy to the thermal conductivity of the phase change heat.

2.1.11. Effect of Space Dimension

The ice point of the water can change from 105 $c^0$to 8K with the change of the nano-pipe diameter, and even the permanent liquid water will be present. This is a very good example of the space dimension effect in phase change. It is easier to understand the influence of spatial dimension on phase transition by discussing these experiments with two kinds of potential energy.

In 2005, Alexander Kolesnikov et al. obtained "permanent liquid water" in nano-tubes at near 0K. In 2017, MIT obtained a freezing point of 105c° in 1.05 nm carbon nano-tube, while the smaller size nano-tube had a lower freezing point, which could be reduced to -83c°. As the diameter becomes smaller, the freezing point can be reduced to 8K, and even the liquid water can be owed forever. These two anomalies in the liquid-solid phase transition of water in nanotube, that is, the total abnormal increase of freezing point in a specific pipe diameter and the water on absolute zero degree is still permanent liquid, which can be explained by dimensional effect.

Water molecules have the ability to form hydrogen bonds, which will be discussed in Chapter 4 as a spontaneous process. However, the capability of influencing the phase transition is limited by the geometrical space, and the formation of three-dimensional hydrogen bond configuration directly influences the freezing point.

6 water molecules can construct plane configuration by hydrogen bonds, and 5 water molecules can construct tetrahedron with a three-dimensional structure. It can be assumed that permanent liquid water can only be a hydrogen bond chain of 2 water molecules.

The most important dimension in the liquid-solid phase transition is the liquid cluster dimension, which is related to l^D. The hydrogen bonding dimension is critical at the cluster scale of the water framed by the hydrogen bonding. The scale of geometry space directly affects the dimension D of the hydrogen bonds and the correlation length l after cluster growth should be affected. And the sufficiently large cluster size makes N in $(V - Nb)$ so large that the space for cluster interaction becomes small that phase transition can occur. However, when the geometric space is not large enough, the clusters can't collide with each other, so internal pressure and phase change can't be generated.

A regular icosahedron with 65 molecules was calculated using a hydrogen bond length of 157.8pm and the circumscribed sphere diameter was approximately 268.46pm, while 19 of these regular icosahedrons packed into the smallest cluster had a circumscribed sphere radius of 795.5 pm. 2 such clusters will be trapped in 1.05nm nanotube. That is V in $V - Nb$, which is determined by the tube diameter, is very small relative to the icosahedron, but N, which is determined by the icosahedron, is very large. There is little space available to resist the rotational momentum and icing can occur at very high temperatures.

When the pipe diameter becomes smaller, V becomes smaller, hydrogen bonds of water can be constructed in a tetrad form, the cluster size becomes smaller, but the N is smaller in the tetrahedral cluster and has larger $V - Nb$, stagnation of rotation momentum is not easy to occur, and the freezing point is reduced along with the reduction of the pipe diameter. The ice lining is already a two-dimensional hydrogen bonded structure, and there is no liquid cluster. The intermediate water molecules surrounded by ice lining could has only a linear linking hydrogen bond, and the liquid can be in the presence of permanent liquid.

The tube diameter is enlarged to 1.4nm, and V is increased. But the space still not enough to put down 2 clusters of regular icosahedrons, the single cluster of regular icosahedrons cannot increase N in proportion, so that there is a larger $V-Nb$, and the freezing point is lowered to 8K.

The $(V-Nb)$ determine the amount of internal pressure that can produce an intermittent phase transition, and thus a transition of different types of momentum $\Delta(TS_B)$ occurs. Without momentum space reconstruction, phase transitions are not mentioned. This scale effect is therefore a best evidence of the existence of two potential energies in thermodynamic systems.

2.1.12. Two Kinds of Phase Transitions

Spatial configurations of molecular translational, rotational and vibration momentum in the discontinuous phase transition undergo abrupt changes, as do corresponding interactions. This is the best evidence that the various momentums have respective two potential energies.

However, in continuous phase transition, the interaction does not change, but the momentum space configuration changes, and the main characteristic is that the molecular interaction changes into cluster interaction. Since Boltzmann entropy can be calculated using N/b. Here b is the number of molecules in the cluster. The successive phase transitions have logarithmic singularities spikes in corresponding physical quantities, such as specific heat logarithmic spikes. It is corresponding to a change count-related physical quantity within the system.

$$\Delta S_B = k_B \ln\left(N/b\right)/N = k_B \ln 1/b \tag{2.8}$$

But some momentum is frozen or released in the discontinuous phase transition. When the phase transition temperature and pressure are kept constant, the molecules in the original phase are always retained, and retained molecular numbers are equal to N_{end}, so that has the following formula .

$$\Delta S_{B,\alpha} = k_B \ln \frac{N_{\alpha,\,end}}{N_\alpha} \to -\infty \qquad \left(N_\alpha >> N_{\alpha,\,end}\right) \tag{2.9}$$

$$\Delta S_{B,\beta} = k_B \ln \frac{N_\beta}{N_{\beta,end}} \to \infty \qquad \left(N_\beta >> N_{\beta,end}\right) \tag{2.10}$$

Infinity merely indicates an order of magnitude difference. The sign indicates only the endothermic or exothermic effect. The difference between continuous and discontinuous phase transitions is the large difference between the number of molecules forming clusters involved in the continuous phase transition and the number of molecules that are transformed into another phase in discontinuous phase change. The discontinuous phase change heat should be similar to the equation (2.4) due to the difference in momentum between molecules and clusters in phase α or β at phase change temperature, which is required by the phase change kinetics. Should be expressed as

$$Q_{Let} = Q_{clu} \approx \Delta S_{B,\alpha\to\alpha,clu} \cdot \Delta T_{\alpha\to\alpha,clu} \approx \Delta S_{B,\beta\to\beta,\,clu} \cdot \Delta T_{\beta\to\beta,clu} \tag{2.11}$$

The kinetic processes in phase transformation are rarely discussed. Phase change heat is generally discussed only as a heat source term or sink term in a continuous medium. Consider its heat conduction process in two phases.

2.1.13. Boltzmann Entropy Change Induced Heat Wave in Cluster Formation

The heat stress corresponding to Boltzmann entropy change during cluster formation generates an inverse temperature gradient and pressure gradient near the phase interface. They are opposite to the pressure $-dp/dz$ and temperature gradients $-\partial T/\partial z$ originally directed at the interface. The z coordinate is the vertical interface direction. Since the heat of phase change is large, the heat-affected region is not limited to clusters. And it can be the region near the phase interface. A mechanical vibration source and a heat vibration source about temperature at the region for the continuous medium can be formed. They can produce a wave process at the level of the continuous medium. The temperature vibration source can generate heat waves, and a heat source such as sinT can generate solitary waves. The thermal wave continuously transfers a portion of the phase change heat to distant molecules, and near the phase interface, more clusters are generated due to heat loss, and the heat wave transport is maintained. This problem is discussed in detail below the section 2.2. The mechanical vibration generates rectification effect on the three-dimensional random momentum of the mass point in the continuous medium and the molecular motion in the cluster. That is, these momentums are combined with the vibration direction of the particle and the vibration direction of molecules in the cluster, and spatial resonance is a kind of spatial degeneracy of the microscopic state. This also generates heat generated by Boltzmann entropy change. Thereby the different distance space between clusters and molecular distance space inside the cluster could form an orderly crystal structure for solidification.

The generation of such mechanical and thermal vibration sources is based on interaction invariant symmetry $p\not{V} = TS_B$. It is only this kinetic process that is sufficient to show that the phase change is completed in one phase. Without TS_B and this symmetry, the phase change problem cannot be understood.

This is so as an important reason why condensed matter physics is of interest to physicists.

2.2. Heat Conduction Wave Equation in Crystal

Maxwell proposed the problem of thermal conduction with limited heat propagation velocity for over 150 years, to date not solved. He believes that the heat conduction equation should be a heat conduction hyperbolic wave equation with finite wave velocity, rather than a parabolic equation with infinite propagation velocity.

The parabolic heat conduction equation is established from the Fourier's constitutive relation of heat conduction in a continuous medium, i.e., the heat flow rate is proportional to the temperature gradient. Temperature gradients and heat flows can be instantaneously established in the entire geometric space, i.e., the heat transfer has an infinite propagation speed.

The main momentum types of gas, liquid and solid in the phase change problem is different. In the heat conduction problem, the translation, rotation and vibration momentum of molecules all contribute to heat transfer; including free electron flow in the metal, also contributes to heat conduction. The regular and flawless crystal was chosen as the target of discussion because only the vibration momentum and Hooke force need be considered in the crystal.

The most used models in the finite thermal wave equation problem are the "phonon" model and the "relaxation" concept. These are all concepts at the molecular level. The thermal problem is fundamentally the one of the molecular momentum levels about change of molecular TS_B potential energy. But temperature is the statistic value of particles in a continuous medium defined by molecular momentum. So the problem on this continuous media level is discussed here based on molecular motion.

2.2.1. Spontaneous Reverse Potential due to Microcosmic State Degeneracy

The phonon model of a crystal is described by a normal mode of the three-dimensional vibration momentum. The vibration mechanical energy in the model propagates in the form of sound waves. But the heat in thermal conduction can't be propagated as acoustic waves. The vibration source of the heat wave propagation is not a phonon vibration source.

The Hook force of mass point giving rise to heat conduction oscillator in a macroscopic continuous medium is created by the rectification and degenerating effect of the vibration momentum of the molecules. The generation of the macroscopic pressure stress in the continuous medium crystal is the result of reconstructing a molecular three-dimensional random vibration momentum field under the action of the temperature difference potential of $\Delta\left(TS_B\right)_{Mac}$ in the macroscopic temperature field. An external thermal potential $\Delta T_{Sys} = T\left(L\right) - T_0$ in the plate system causes the macroscopic state of the system in the high temperature region to change from the $S_{B,Sys} = 2$ temperature state to the $S_{B,Sys} = 1$ with the same temperature state. Assuming that from T_0, reaching an average temperature T for a localized heat-affected zone $A\left(\Delta T_{Sys}\right)$ that change number of ensembles ΔS_{ens} within the local region and thus $S_{Sys}\Delta T_{Sys} = T_0 \Delta S_{ens}$.

The number of ensamples affected by thermal potentials is W_{ens} in the process, so there is $\Delta S_{B,ens} = k'_{B,ens} \ln(1/W_{ens}) << 0$ in the local zone $A\left(\Delta T_{Sys}\right)$ affected by external heat source. Here is $k'_B = k_B / b$ that b is the number of molecules defining an ensemble. In the A domain under the action of the temperature difference potential ΔT, all the ensemble numbers W_{ens} with random vibration momentum are degenerated. This microscopic degeneracy process will therefore produce an amplified reverse thermal potential $S_{B,Sys}\Delta T_{Sys} << 0$, giving off heat and resulting in an inverse temperature differential potential $-\Delta T$. Heat will flow in the reverse direction to the high temperature zone, creating a reverse heat flow. This temperature oscillator is the primary reason of the wave propagation of heat between the macroscopic particles. Based on the principle of interaction invariance the formula corresponding to this macroscopic thermal fluctuation is shown in the specification

$$T_0 \Delta S_{B,ens} = \not{V}_{ens} \Delta p_{ens} = \not{V}_A p_{Sys} << 0 \qquad \left(A \in A\left(\Delta T_{Sys} \right) \right) \tag{2.12}$$

I.e., $\Delta p_{ens} < 0$ and $p_{Sys} < 0$ will be generated in the zone affected by heat. As opposed to p , that original p can rectify the weaker vibration momentum by a strong vibration momentum. These two alternately appearing forces generate oscillator of mass point, which gives rise to vibration of mass points. This is also the Hook force between ensembles that can result in temperature vibration in the thermal oscillator due to degenerating process and spread out the heat from the localized heat source in form of waves. In other words, a heat wave is a process degenerating random vibration momentum of ensemble when it is gradually advancing. In response to this macroscopic thermal wave propagation process, the force of this rectifier and degeneracy effect according to $\left(p\not{V} \right)_{\Delta T} = \left(TS_B \right)_{\Delta T}$ is a macroscopic Hook force of mass point. And simultaneously form a pressure stress field which generates macroscopic vibration in the crystal. This thermal stress was visible in photo elastic experiments. When this macroscopic Hook force produces particle vibrations, it also produces rectifying and degenerating effect on the microscopic Hook force between molecules in clusters. There are $W = b$ molecules in an ensemble. By means of $\Delta S_{B,ens} = k_B \ln 1/W$, the random molecule vibration momentum after rectification and degeneracy effect will also heat flux release, which will have a temperature gradient between the particles of the continuous medium. And the temperature on the crystal body gradually increases to the final state. That is, rectification and degeneracy effects arising spontaneous reverse potential are generated on both levels. The latter is also an irreversible reason of thermal conduction under limited temperature differential.

From the molecular motion level, the Hooke force between molecules generates sound waves, but does not generate macroscopic heat waves. The acoustic wave generated in the crystal by the mechanical vibration source is a linear wave. The heat source generates heat wave, which is a kind of dispersion wave. Because the potential energy TS_B is changed due to dT change, the match fraction $\delta n_i\left(\delta \bar{V}_i\right)$ of molecular momentum is changed, and $\delta \bar{V}_i$ is changed by Hooke force in the middle, so that linear direct response cannot be realized; the key point is that the generation of macroscopic heat flow can be generated undergo the degeneracy of microscopic states for vibration momentum. Macroscopic heat conduction cannot be a linear wave. Spatial redistribution of particle number during population requires time and a rebalancing process that transfers momentum at infinite speed is not possible.

The rectification degeneracy effect for original microscopic motion resulting in macroscopic wave behavior under external potential can be a universal mechanism. Perhaps this is a truly important link between microscopic random motion and macroscopic deterministic motion. The importance of which is far more important than the problem of heat conduction.

2.2.2. Heat Wave Equation

The heat transfer problem is defined herein as: an energy local input area L (t) which forms a heat source with a temperature higher than that of the surrounding area and may not be a constant temperature heat source; the problem of spreading its energy in the form of "heat" to the surroundings. This kinetic process can therefore be discussed on a macroscopic continuous medium level.

In a regular ensemble discussion, one unit in phase space is one of the specimen ensembles, and the temperature of the constant temperature heat source is different from the other specimen ensembles having the same temperature. By using this phase space concept, it is possible to be compatible with the concept of distance space. Sufficient molecules per ensemble, which can define the temperature, are "bridges" between the molecular motion and the motion of ensemble or mass point in continuous medium.

After the thermodynamic equilibrium of the system is broken, the heat transfer process is the kinetic process of the system re-equilibrium. The difference between $\delta\mathcal{V}_i$ in this region and $\delta\mathcal{V}_{i,0}$ of the original equilibrium state results in a temperature difference $T(l)-T_0$ in the region, thus defining the momentum space configuration potential as $T\iiint_{\mathcal{V}_{ens}} dn_i(\delta\mathcal{V}_i)$ for $\mathcal{V}_{ens}$ distance space of an ensemble. The sample subsystem at the high temperature region and other sample ensembles can change under the drive of the temperature difference $\Delta T\iiint_{\mathcal{V}_L} dn_i(d\mathcal{V}_i)+T\Delta\iiint_{L} dn_i(d\mathcal{V}_i)$. The expression means the spatial diffusing process of the rebalancing process of the vibration momentum of the energy input area on the ensemble level due to the driving action by means of the temperature difference. The meaning of the expression is a space diffusion process of the rebalancing process for the vibration momentum in the area input energy due to the driving of the temperature difference between the ensembles.

On a macroscopic level, Boltzmann entropy is reduced spontaneously according to the second law of thermodynamics

$$\Delta S_B = S_B(T)-\left[S_{B,0}+S_{B,T_1}\left(L(t)\right)\right]<0 \tag{2.13}$$

The rebalanced system will tend to be the same temperature T. Because there are enough particles in the subsamples of the regular ensemble, they can be reassigned with $\delta\mathcal{V}_i$ changes. Its macro dynamic behavior is to obtain the time variation of statistical value at the position of constant geometric space position, again to be put back to Euler coordinate in crystal, to use the temperature defined by regular ensemble to discuss the heat transfer problem in continuous medium, that is to say, there is

$$\frac{D(TS_B)}{Dt} = S_B \frac{\partial T}{\partial t} + C \cdot \nabla (TS_B) \tag{2.14}$$

This is a representation of the macroscopic heat flow at a fixed space point in the Euler absolute space-time coordinates. The lattice can't flow so and the flow for the spatial coordinates can only be heat, C is the finite heat propagation velocity to be sought. To obtain the macroscopic kinetic equation of heat transfer, the corresponding Hamilton function needs to be constructed.

The definition of kinetic energy is different for different motion levels. It is Molecular kinetic energy in a microscopic system, and kinetic energy in macroscopic thermodynamics is the average temperature in an ensemble; the local temperature may have a spatial gradient. The temperature is a statistical average of molecular kinetic energy and does not contain time variable. But when the system is unbalanced, the rebalancing process varies with time. The kinetic energy of the thermal oscillator in macroscopic statistical significance needs to be defined in $\partial T/\partial t$ of above (2.14), and is the macroscopic time change rate of vibration kinetic energy. The corresponding kinetic energy function can be noted as

$$T(t) = \frac{1}{2} l(t)^6 \left(\rho c_v \frac{\partial T}{\partial t} \right)^2 = \frac{1}{2} l(t)^6 (n'mc_v \frac{\partial T}{\partial t})^2 \tag{2.15}$$

The macroscopic density ρ is defined by the ensemble sample, the macroscopic temperature T is defined by the ensemble sample, n' is the number density of particles in the ensemble sample of the thermal vibrator, here m is the mass of lattice particles (atoms or ions), and c_v is the constant specific heat capacity. Here $l(t)^3$ is the phase volume element of the ensemble sample.

The spatial distribution potential energy $T\square\iiint_{V_{ens}} dn_i\left(dV_i\right)$ of macroscopic momentum for the thermal vibrator should of course be in the form $\nabla\left(TS_B\right)$ of a macroscopic spatial gradient defined by the ensemble sample. The potential function of the thermal vibrator is recorded as

$$V=\frac{1}{2}[kn'l(t)^3 l(t)\nabla T]^2=\frac{1}{2}l(t)^8[n'k\nabla T]^2 \tag{2.16}$$

That is used the spatial gradient of the general heat flux density by using Fourier law $q=k\frac{\partial T}{\partial n}$. Among which k is the thermal conductivity of the medium. The Hamilton function of a thermal vibrator can be described as

$$H=\frac{1}{2}l\{t\}^6(n'mc_v\frac{\partial T}{\partial t})^2+\frac{1}{2}l(t)^8(n'k)^2(\nabla T)^2 \tag{2.17}$$

Use $\alpha^2=\left(n'mc_v\right)^2$ and $\beta=n'k$ rewrite the above function. The Euler equation in the absolute time space is obtained according to the Hamilton principle and by finding the variation of the Hamilton function, namely

$$\frac{\partial H}{\partial T}-\frac{\partial}{\partial x_i}(\frac{\partial H}{\partial T_{x_i}})-\frac{\partial}{\partial t}(\frac{\partial H}{\partial T_t})=0 \tag{2.18}$$

The vibration equation of the thermal vibrator with time-varying mass can be obtained

$$\frac{\partial^2 T}{\partial^2 t}+\frac{6}{l(t)}\frac{dl(t)}{dt}\frac{\partial T}{\partial t}=l(t)^2(\frac{\beta}{\alpha})^2(\frac{\partial^2 T}{\partial x_i^2}) \qquad (i=1,2,3) \tag{2.19}$$

Is rewritten into

$$\frac{\partial^2 T}{\partial t^2} = c_0^2 \frac{\partial^2 T}{\partial x_i^2} - 2b\frac{\partial T}{\partial t} \qquad |(i = 1,2,3) \tag{2.20}$$

Here $c_0 = \omega_0 l_0$, $\omega_0 = \frac{\beta}{\alpha} = \frac{n'k}{\rho c_v}$, $b = \frac{3}{l(t)}\frac{dl(t)}{dt} = \frac{3u(t)}{l(t)}$ the initial wave velocity and the initial frequency of the wave propagation of the heat flow generated by the thermal vibrator are respectively, and b the damping coefficient of the thermal vibrator. Here $u(t)$ is the change speed of the ensemble phase element scale. This equation is generic and also applies to fluids. Because the molecular vibration momentum in the gas and the liquid can complete the rearrangement of the molecular number for same macroscopic energy level reconfiguration in a heat conduction mode. This will discuss further in the next chapter about the thermo-acoustic effect and vortex tube R-H effect in the gas. The following problem is to be discussed.

2.2.3. Thermal Conduction with Limited Wave Velocity

A thin plate with thickness z, width b and temperature uniformity at b and z direction, under adiabatic condition, leading edge l_0 is contacted with heat source $T(0)$ generate a thermal vibration source with macro scale as following boundary and initial condition

$$\begin{aligned}
&T(0) = T_1(0, l_0(0)) \\
&T(0) = T_0(0, l) \\
&Q_0 = \rho c_v bz\left[T_1(0, l_0(0)) - T_0\right] = \int_0^t dt \int \rho c_v \left[T(t, l) - T_0\right] zbdl = \\
&\quad = \rho c_v bz l_b \left[T_2(\infty, l_b) - T_0\right]
\end{aligned} \tag{2.21}$$

Because of the circular symmetry and the uniform temperature of the thin plate on z coordinates, there is a general one dimension solution

$$T = T_1(0)e^{i(kl-\omega t)} \tag{2.22}$$

Substituting into equation to obtain frequency, wave number and wave velocity

$$\begin{aligned} \omega &= \pm\sqrt{c_0^2k^2 - \frac{9u^2(t)}{l^2(t)}} + 3\frac{u(t)}{l(t)} \\ c &= \frac{\omega}{k} = \pm\sqrt{c_0^2 - \frac{b^2}{k^2}} + 3\frac{b}{k} \end{aligned} \tag{2.23}$$

The frequency in the solution is time-varying, so the wave velocity is not a constant or linear wave at each instant, but a dispersion-like wave. Moving by group velocity of wave packet

$$C_g(t) = \frac{d\omega}{dk} = \pm\frac{kc_0^2}{\sqrt{c_0^2k^2 - b^2(t)}} \tag{2.24}$$

The heat flow is in the form of wave packets, spreading at group velocities. The thermal conduction is propagated at a finite wave speed. It is considered that $k = 2\pi/\lambda = 2\pi/l(t)$, when $c_0^2k^2 > b^2$ there is a finite wave velocity. And the wave velocity becomes infinite while $c_0^2k^2 < b^2$. However, when $c_0^2k^2 = b^2$, that is $u(t) = \frac{2\pi}{3}\frac{n'k}{\rho c_v l_0}$. What is characterized the moving speed of the area of thermo-oscillator that depends on the asymmetry degree of the kinetic energy and the potential energy of the momentum space configuration in the thermo-oscillator. Before this velocity increases to an infinite wave velocity, the conductive diffusion still has a finite wave velocity.

2.2.4. Damping Effect

The heat wave propagation mode is affected due to the change of the size of the thermal vibrator. When the rate of change of kinetic energy in the thermo-oscillator is slower than the change of the thermo-oscillator's own scale, the relation can be gotten on the magnitude scale analysis

$$\frac{\partial^2 T}{\partial t^2} \sim \frac{u}{l}\frac{\partial T}{\partial t} \Rightarrow 1 \sim \frac{\delta l}{l} \tag{2.25}$$

When the scale change of the thermal vibrator is equivalent to the original scale, the change of kinetic energy in the thermal vibrator is slower than the energy rebalancing of thermal potential energy, namely the scale change of a heat affected zone is very fast and exceeds the size of the original thermal vibrator zone. When this $\delta l >> l$ happens, the equation becomes to the classical thermal conductivity equation.

$$\frac{\partial T}{\partial t} = \frac{c_0^2}{2b}\frac{\partial^2 T}{\partial x_i^2} \tag{2.26}$$

The time of change δt is implied in the above quantitative analysis. Over short time intervals, this case $\delta l >> l$ is always true for very short distances. However, this is a problem that $\delta l << l$ case occurs at long time intervals over a very long distance. The damping term is a small term, and equation (2.20) changes a standard hyperbolic heat wave equation.

$$\frac{\partial^2 T}{\partial t^2} = c_0^2 \frac{\partial^2 T}{\partial x_i^2} \qquad (i = 1, 2, 3) \tag{2.27}$$

2.2.5. Solitary Wave

The solitary wave based on the hyperbolic wave equation above have an additional energy source or sink term, with similar $(\rho c_v l_0^3)^2 \sin T = m^2 \sin T$ source terms, which may be added to equation (2.27). This is the exogenously input energy and or the potential energy TS_B of its own momentum redistribution nature. So it should appear in the equation, i.e., noted as not being in the boundary condition or initial condition

$$c_0^2 \frac{\partial^2 T}{\partial x_i^2} - \frac{\partial^2 T}{\partial t^2} = m^2 \sin T \quad (i = 1, 2, 3) \tag{2.28}$$

This is a Sine-Gordon Equation, which may have solitary wave solutions.[15]

2.2.6. Vibration Solution

The Euler variation in the absolute time space, TS_B adopts a gradient expression form, and the heat conduction equation is only constructed. For the distance space where the vibration momentum interacts, the linear acting force term $\omega_0^2 T(t)$ can also be constructed by the temperature changing along with the time, the thermal vibrator vibration equation in the acting distance of the thermal vibrator is written as ,

$$\frac{\partial^2 T}{\partial t^2} + 2b \frac{\partial T}{\partial t} + \omega_0^2 T = 0 \tag{2.29}$$

Equation (2.18) the vibration frequency of the thermo-oscillator describes the vibration of the thermo-oscillator, relative to the rebalanced temperature T_2, which occurs with amplitude between the temperature T_0 of the thermo-oscillator and the surrounding crystal that is also constantly damped

$$\omega = \pm\sqrt{\omega_0^2 - b^2} - bi \tag{2.30}$$

When $\omega_0^2 \le b^2$, an over-damping or critical damping process appears. The thermo-oscillator temperature monotonically decreases to T_2. As it is commonly observed heat conduction in solid.

2.3. Sources of Irreversibility and Frictional Heat

The Clausius entropy proposed has the concept of irreversible processes, and due to the "Theory of Hot Silence" the irreversible problem becomes too large and must be clarified. Then there are various "irreversible" sources of discussion. The Clausius inequality is the criterion for the cyclic process, and inapplicability in the most general arbitrary process. Therefore, it is not a true irreversible criterion. But after Boltzmann entropy proposed, the irreversible problem does not make clearer.

2.3.1. Friction Generate Heat

Heat conduction, mixing and friction heat are three well-recognized classical irreversible processes. The irreversibility of the two processes of heat conduction and mixing is still reversible by virtue of infinitesimal temperature difference and concentration difference. However, the irreversibility of friction generate heat is not well explained, there appears to be no description of infinitesimal friction, and it is an absolutely irreversible process. Therefore, understanding the irreversible nature of friction is helpful to clear the knowledge of the source of the irreversibility.

2.3.2. Friction Generate Heat on the Solid Surface

Joule proves that the heat-work equivalent test is for the frictional heat generated by mechanic work. However, no one has explored the microscopic mechanism. The work done on the system is always greater than the momentum increment gained by the system. It is believed that, for example, when a solid body moves over a solid surface, the solid contact surface has frictional resistance and consumes the work input.

The solid is driven to move by macroscopic acting force, and the momentum increment of the macroscopic motion of the solid is generated. But the interaction force in the solid is the Hooke force within molecules, and the motion of the solid molecules is the vibration momentum to correspond the friction heat. These two movements are completely independent motion at two levels of the interacting forces. But both create a bond at the solid contact surface. Friction is a resistance to macroscopic thrust, coupled with macroscopic motion. The friction is microscopically opposing intermolecular Hooke force on the solid surface layer and can lock in molecular vibration momentum, which thus also retards molecular motion and generates heat. On a microscopic level, however, such microscopic friction forces are localized and random in space and time. This randomly occurring friction force on a localized area of the friction surface is a molecular level interaction sufficient to lock in the molecular vibration momentum on these interfaces. By this relation $\Delta\left(f\,\delta A_i\right) = \Delta\left(TS_B\right)$, it can be interpreted that the friction f is on a random area within the range δA_i of the distance space that can be acted upon. The molecules with the locked vibration momentum are equivalent to the molecular vibration momentum reduced in the ΔS_B number of the molecules, and corresponding heat $T\Delta S_B$ is released according to the average vibration kinetic energy of the molecules. The frictional heat is generated by the locking of random local solid molecule vibration momentum under the macroscopic thrust force, that is $\Delta(TS_B)$.

2.3.3. Friction Heat in Fluid

The flow of fluids is a macroscopic translational motion driven by pressure in a continuous media. The resistance on the solid wall generates macroscopic shear force, and fluid velocity on the wall is equal zero under the condition of no slip, namely the fluid rotation momentum is stopped. As described above, frictional heat is also generated. However, from the microscopic rotational momentum of the molecules, all molecules on the wall are not entirely arrested, since adjacent fluid molecules can exchange rotational momentum through very close elastic collisions, and thus the molecular rotational momentum is arrested, both in terms of position and time, randomly. Although the vortices defined for the mass point are all equal to zero. In addition, the fluid shear force defined by continuous medium mechanics can generate vorticity, namely angular momentum defined by macroscopic shear displacement. But also gather into macroscopic concentrated vorticity in the local random negative pressure gradient region. (Concentration vortex will be discussed in the next chapter). The concentration vortex has a co-located rotation center only when the concentration vortex combines a large number of infinitesimal spin vorticity to revolve that is the isospin motion. This vorticity reduction process ΔS_B also releases heat. The vorticity of molecule or particle defined by the ensemble is the spin motion around the center of mass. The randomness of such macroscopic concentrated vorticity manifests itself as local and spatial randomness. The heat generated by friction therefore has no temporal and spatial translational invariance, i.e., the process has no temporal and spatial translational symmetry. There is no conservation of energy and momentum. The source of the irreversibility is the spatial-temporal translational symmetry breaking that is caused by the random motion occurrence on microscopic level within the macroscopic motion. And it is not derived from asymmetry of time reversal for the kinetic equation.

The problem of friction of the gas flow is only to illustrate the translational motion in the continuous medium, and the micro translational motion of molecules is also included. The friction heat on the solid wall surface is also generated by stagnation of molecular translation kinetic energy.

2.3.4. Heat Transfer Limited Temperature Differential

It is another classical irreversible process. Again, taking solid thermal conduction as an example, the microscopic dynamics of the molecular vibration momentum in the process and the equation for heat transfer at the level of the continuous medium were discussed above. The spontaneity of Boltzmann entropy decreasing and the function of the ΔTS_B in the aspect of constructing a heat conduction equation are mainly discussed.

The relationship between macroscopic and microscopic levels in thermal conduction is still a clue to explain the irreversible process. The heat conduction problem of two heat-insulating gas containers with different temperatures on the intermediate partition board is that the gas in the containers has no work exchange process in the process of reaching the equilibrium temperature after the heat exchange in the isometric process. Its Clausius entropy change is calculated thermally according to an reversible isometric process. The Boltzmann entropy change of the continuous medium macroscopic layer can be counted as 2 thermodynamic macroscopic states according to the condition that the temperatures for two objects are respectively T_1 and T_2, and the macro state sum equal to 2 before heat conduction and the temperatures T is counted as 1 macroscopic state after heat conduction. The Boltzmann entropy of this process thus becomes

$$\Delta S_{B,Mac} = 1 - 2 < 0 \tag{2.31}$$

The calculation of the Boltzmann heat conduction entropy is suitable for irreversible or reversible heat conduction processes. It is also applicable to the heat conduction process of ideal gas, different from the $\Delta S_{C,Mac} = C_v \ln T_1/T_2 > 0$ calculation using reversible isometric process. But when $T_1 \to T_2$, $\Delta S_{C,Mac} \to 0$ and $\Delta S_{B,Mac} \to 0$. The two entropies are calculated identically, i.e., the heat conduction process of the infinitesimal temperature difference is reversible. For motions of different levels there is typically momentum at different levels and $p\cancel{V} = TS_B$ symmetry under corresponding interaction. In the heat conduction process, pressure changes defined by continuous medium mechanics are arranged in two containers according to a gas state equation and an isometric process

$$p\cancel{V} = TS_B \quad dQ = Vdp \tag{2.32}$$

From a $\Delta(TS_B)_{Max} = \Delta(p\cancel{V})_{Max}$ symmetry point of view, there should also be macroscopic interactions dp_{Max} corresponding $\Delta S_{B,Max}$ on the continuous medium level for heat conduction process. The change in pressure at the macroscopic level of this limited heat transfer differential is opposite in the two vessels. The reconstruction of the spatial distribution of the respective microscopic molecular momentum is a resistance. Such as the high momentum molecules transfer momentum to the lower temperature vessel where the pressure is increasing and the low momentum molecules absorb momentum from the higher temperature vessel where the pressure is decreasing.

The pressure generated by elastic collisions at the molecular level is the force of the rebalancing of the gas translational momentum. Finally, the temperature of the two objects is allowed to approach the same temperature. In thermodynamic systems with limited temperature differences, intermolecular forces are random for macroscopic determinism. Boltzmann entropy change of ideal gas-to-gas heat conduction process in two insulated containers for microscopic molecular motion is calculated according to the distinguishable locator for different distance space δV_i and the randomness of indistinguishable local off domain statistics calculation of identical particles for the same δV_i. Randomness in the conduction of heat is thus a constitutive characteristic of the system. For macroscopic deterministic interactions dp_{Max}, random motion is used to achieve rebalancing momentum configuration during heat conduction. This microscopic stochastic action in macroscopic certainty is the source of irreversibility. The heat conduction process is reversible at $dp_{Max} \to 0$ times as well $dp \to 0$.

2.3.5. Mixing Process

The box space with the wall, after mixing of two kind of gas, the final equilibrium state achieved, and required the homogeneity of not only, the material but also the momentum. The macroscopic mass flow occurring in the process is the isothermal and isobaric mixing process. For each gas, it is a process from total pressure drop to partial pressure process in a process of volume expansion. The mixing process thus has macroscopic pressure drop $dp_{A,Max}$ and $dp_{B,Max}$. And it's all less than zero. The macroscopic Boltzmann entropy for mixing process macro state should be

$$\Delta S_{B,Mix,Max} = 1 - 2 < 0 \tag{2.33}$$

From the symmetry of $\Delta\left(TS_{B,Mix}\right)_{Max} = \Delta\left(pV\right)_{Mix,Max}$ there should also be $dp_{A,Max}$ and $dp_{B,Max}$ or mixing process.

The Boltzmann entropy of an ideal gas molecule mixing process is a statistical result of the translational momentum of two kinds of gas molecules under the molecular interaction force p. There is thus randomness in the off-domain statistics. The co-action of the macro determinacy of $dp_{A,Max}$ and $dp_{B,Max}$ with the intermolecular force p on the random partition of the same energy level molecules in the corresponding distance space should be an irreversible source in mixing process. The thermal conduction and mixing may be reversible under infinitesimal macroscopic action conditions. But friction generate heat has no reversible process. Since the random resistance also occurs on a macroscopic level.

But this result is not necessarily generalized because it has to be based on a multi-body statistical system. It is not known whether the result can apply to single particles statistical system with spatial distribution probability of the momentum. It should be questionable whether irreversible is a universal question.

But the living body is constructed of a macroscopically continuous medium. This irreversible discussion is applicable to the life activities. The order of life is not equal to the state where the living system itself is not far from equilibrium, which is in a thermodynamic equilibrium system that can be described by a grand canonical ensemble with heat and mass exchange under infinitesimal temperature differences and chemical potential differences. However, the living system exists in an environment of the earth with limited temperature difference, concentration difference and chemical potential difference, and an irreversible process of metabolism is continuously performed.

The irreversible processes of thermodynamic systems derive only from the randomness in the macroscopic action of the system.

Discussion of irreversible sources implies-very important suggestion: the processes that actually occur are not a single one but are one of combining motions at different levels. The interactions decomposed into respective motion level are analyzed, and each-interaction should obey energy conservation and momentum conservation. Otherwise the question is translational symmetry for time and space. Irreversible is therefore only a result of entanglement of different interactions. Symmetry is the bottom line that physics must adhere to.

Chapter 3

PROBLEMS OF MOTION ON CONTINUOUS MEDIA LEVEL

The surrounding world, including the life itself, is presented usually in the form of a continuous medium. They exist in the form of congregating large number of molecules or atoms. This chapter discusses the problem of macroscopic bulk motion of such congregate matter. They are related to mechanics that is modern scientific provenance. This is also the most mathematically deterministic science developed via Galileo, Newton, Lagrange and Hamilton. But also leaves a problem that is not solved so far, not just the mathematical uncertainty brought by the non-linear problem. The problem of turbulence in the fluid is an example, and the fact that has become increasingly clear in recent years is: the vortex is a bi-directional cascading process. We know how to describe the vorticity of a macroscopic mass point by means of a strict mathematical definition, which is the spin motion of a particle or an ensemble. It is not visible to the naked eye. But we still don't know how to define macroscopically visible concentrated vortex in a non-viscous global translation motion. It should be at least a mathematical expression with physical meaning. This will provide a new basis for overall fluid mechanics.

The mass point defined in the continuous medium should be: the smallest set of molecules that can be statistically averaged for all molecular momentum within mass point. Continuous media typically has a fluid and a solid. The fluid includes both gas and liquid. Continuous media mechanics discusses macroscopic kinetic behavior. Classical thermodynamics starts with a discussion of the energy conversion of a continuous medium. This chapter attempts to discuss the conversion between these energies, starting from the macroscopic kinetic energy, the two potential energies and the corresponding thermodynamic expression on the continuous medium level. When the expression method of free energy is followed, it is agreed that it is different from the definition on the molecular motion level. Since the number of particles in a continuous medium, such as a fluid, is far less than the number of molecules, but still sufficient to form a statistical system for an ensemble system. The continuous medium can be the subject of thermodynamic discussion. Its interactions are still a set of interactions in the system rather than individual interactions. The kinetic energy should also correspond to the macroscopic velocity, the corresponding macroscopic translational momentum, the vorticity and the vibration momentum. The corresponding forces are matched to the "distance space" of the macroscopic motion. The translational velocity of the ideal fluid may be a rod-like translation relying on pressure differential between two ends of the rod. The fluid is not able to withstand any shearing action and distances space scale l in $\mathrm{Re} = ul/\nu = 1$ can produce macroscopic micro-beams, but also vortex at large space scales can be to several thousand meters like typhoons. Of course, the pressure, shear as well as Hook stress are interactions between the macroscopic particle momentum changes. The complete set of equations for continuous medium gas dynamics must include the gas state equation and adiabatic relationships, and also shows that this $pV = TS_B$ symmetry between momentum and energy still holds at the continuous medium level. The gas state equation is originally generated from the ideal gas of the continuous medium. Therefore, the most significant new concept in continuous medium mechanics can also be introduced, and the deviation of a thermodynamic equilibrium state in a kinetic system not only depends on the potential energy TS_B for restoring rebalance to reach a new equilibrium state, but also can recombine various momentum combinations to generate new forms of movement. Although we are used to attribute the energy loss of the flow to viscous shear dissipation, no slip boundary condition on the solid boundary shows the ability of TS_B to reconstruct motion types. The TS_B not only stagnate fluid whole translational momentum and generate frictional heat; but also generates macroscopic particle spin momentum, i.e., vorticity in the fluid. This also means the reconstruction of different momentum types in space. Not only is dissipation on a macroscopic level, but it also means that there is a spontaneous process directed according to the second law of thermodynamics.

The only problem here is how defined the microscopic states in the macroscopic classical mechanics? It is believed that particles in the same "distance space" all have the same forces and define the same continuum medium. A "distance space" on a macro scale may define a corresponding macro-state. The thermodynamic statistics of a continuous medium can be carried out using a Gibbs ensemble, as long as there is a distance space distribution under some action distribution.

Among the problems that cannot be described in continuous media mechanics, the mathematical description of concentrated vortex is a very disturbing problem. Although it profoundly affects human survival and daily life in nature, it is so clearly visible; TS_B potential energy on a molecular motion layer is considered as heat dissipation in a basic equation N-S equation of hydrodynamics by viscous term, but lacks the momentum $TS_{B,\,Mac}$ change on the continuous medium level, so the N-S equation may not be complete; but the N-S equation puts the momentum of the different layers together for processing. Even if the strongest computer is used for numerical calculation of the whole N-S equation, we can't still get the concentrated vortexes in good agreement with a coherent structure in the boundary layer experiment results [1]. The problem of turbulence only depends on an experimental model based on the cascade dissipation of the large vortex, and engineering calculation will be difficult to completely solve. The experiment of wind tunnel and water tunnel is needed to enter sea and to go up to the sky. Not to mention explaining of the dual cascade process in atmospheric science [2].

This chapter discusses 3 problems: 1. the definition of concentrated vortices and the sources of excess kinetic energy and heat therein 2. Total temperature separation of R-H effect occurring in vortex tubes generates the refrigeration effect. This is one kind of the heat transport in the form of thermal solitary wave based on the action of pressure arising vibration in the concentrated vortices that output corresponding to the vibration momentum and kinetic energy. It is also the basis for the thermo-acoustic effect and pulse tube refrigeration. 3. The energy source of damage caused by resonance of a solid structural member.

The first problem is that: what are the concentrated vortexes? The discussion of concentrated vortexes can help us to clarify the difficulties presented by missing $TS_{B,\ Mac}$ on continuous media level.

3.1. Mathematical Definition of Concentrated Vortex

The fluids mechanics discusses the problem of the spatial distribution of macroscopic motion velocities of fluid particles evolving over time. The flow field is a momentum field. A stationary fluid has only a balanced force field. Therefore, understanding the momentum field of macroscopic motion for such fluid particles in this sense is only interaction and the thermodynamic equilibrium state in the macroscopic sense. There is therefore no need in hydrodynamics to be based on the assumption of near equilibrium states at the level of molecular motion. In fact, the momentum of mass point translation, rotation and vibration in the high supersonic flow, although may be very large, but it is still a thermodynamic equilibrium state of macroscopic motion. The normal and shear stresses between enormous amounts of the mass point constitute the macroscopic interaction field of the system that corresponds to the momentum field of the macroscopic motion.

There is also the macroscopic symmetry, $pV_{Mac} = TS_{B,Mac}$, it is one interaction that governs the two potential energies in the continuous media systems. Subscripts *Mac* refer to macroscopic continuous media systems.

Although the N-S equation can be well solved for translational momentum fields, propagation of vibration momentum in acoustic and shock waves in compressible fluids, including water hammer, hydraulic jump in liquids, in the N-S equation, despite shear stress term corresponds to the rotation momentum of particle, i.e., the particle spin motion, and it can describe the dissipation process from large vortex breakdown to heat. We also supplemented the N-S equation with a standard turbulence "formation" model. But there is the absence of the concept of $TS_{B,Mac}$ potential energy at the macro continuous media level in fluid motion. It is impossible to establish the relationship between the particle vorticity and the macro visible concentrated vortex. This difference between them is that the particle vorticity is the spin momentum around the mass center axis, while the rotation motion of the concentrated vortex around the common axis in the vortex is similar to the isospin. Under the condition of conservation of angular momentum, the vortex is concentrated in the macroscopic translation motion caused by pressure through the spontaneous process of Boltzmann entropy reduction of particle vorticity. This is the so-called reverse cascade of the two-way cascade (Split energy cascade) proposed by Kraichman [3].

3.1.1. Dynamics and Kinematics of Particle Vorticity

The vorticity equation is obtained by taking the curl on N-S equation that is one in absolute time space system about Euler coordinate.

$$\frac{d\vec{\omega}}{dt} = (\vec{\omega}\cdot\nabla)\vec{V} - \vec{\omega}(\nabla\cdot\vec{V}) + \nabla\times\vec{f} - \nabla\left(\frac{1}{\rho}\right)\times\nabla p + \nu\nabla^2\vec{\omega} \qquad (3.1)$$

The first term on the right of the equation is the vorticity change produced by the stretching and bending action of the velocity gradient on the vorticity axis of mass point. And the second term is the vorticity change that occurs when the particle volume changes under compressible flow. Neither term is the actual interaction force in the flow field. The last three terms on the right of this equation indicate that non-conservative force, non-positive pressure fluid and viscous forces are the three forces that generate particle vorticity. In contrast, without these three actions, particle vorticity flux with body is conserved (Kelvin theorem) according to the kinematics of particle vorticity. In addition, there is still Lagrange theorem about the conservatism of concentrated vortex as well as Helmholtz theorem about strength retention of vortex tube and vortex line. When the three acting forces are not applied in the flow, the mass points of fluid without vorticity are kept to be non-vorticity constantly, and the mass points with vorticity are also kept to be their vorticity constantly in the movement.

The rotational momentum of the molecular level renders the fluid deformable, i.e., macroscopic viscous property, and appears to be unable to withstand arbitrarily small shear stresses at the continuous medium. The viscosity not only generates the vorticity, but also diffuses the vorticity and dissipates the vorticity. Thus, the discussion of concentrated vortices focuses on viscosity that seems to make sense, but the generation of concentrated vortices is just independent of fluid viscosity and shear stress. This is because the accumulation of particle vorticity can be completed in particle translational momentum field to form concentrated vortexes, which will also contain non-vorticity and vorticity fluid particles, in which the dominant force is pressure in continuous matter. The vorticity is conservation only in the flow field with this pressure. Instead, the viscous shear force makes the angular momentum of the mass point in the flow field no longer conservation.

By erasing the shear stress outside the grid boundary, the large vortex simulation numerical algorithm can better simulate the generation of large concentrated vortex. On the basis of cascade dissipation process experiments of vortex breaking, the constructed turbulence model cannot describe a reverse cascade process of the collection of particle vorticity generating concentrated vortexes.

3.1.2. Concentrated Vortex

Under the conditions of non-viscosity, positive pressure fluid and conservative force, the concentrated vortex or particles vorticity are the conservative of vorticity. The question is why can assemblage and degeneracy of vorticity occur. The reason for this degeneracy is that the difference between the state number of mass points with vorticity and the numbers of vorticity in the concentrated vortex is very large. There is only one common revolution axis that all $N_{\omega,\,M}$ particles in the concentrated vortex have only one vorticity value. It means that there is a spontaneous Boltzmann entropy decrease process. Each particle vorticity has its own direction and magnitude, and there is only one macro state number of vorticity for every particle with vorticity. The concentrated vortex has only-one macroscopic state for common revolution axis and the same vorticity value. And thus has

$$\Delta S_{B,\,\omega,\,Mac} = k'\ln\left(1/N_{\omega,M}\right) << 0 \qquad \left(k' = k_b\,/\,b\right) \tag{3.2}$$

Here b is the molecule number in a mass point. According to the principle of force invariance about $pV_{Mac} = TS_{B,\omega,\,Mac}$, the process of concentrated vortex generation is the translation migration process of particle vorticity. The force should still be the pressure that produces the translation momentum of the particle. Therefore, under the action of pressure the $TS_{B,\omega,\,Mac}$ potential energy completes the collection of vorticity and reconstructs vorticity field to form concentrated vortex. The pressure field is analyzed to find the assemblage condition of particle vorticity that should help to find for the condition generating the spontaneous degeneracy vorticity process.

3.1.3. Kinematic Properties of Concentrated Vortex

Because of conservation, the vorticity field generated by the concentrated vortex is a field without a source due to the presence of

$$\nabla \cdot \vec{\omega} = \nabla \cdot \left(\nabla \times \vec{V} \right) = 0 \tag{3.3}$$

In the concentrated vortex motion, the vortex flux must remain unchanged when the cross section is deformed, i.e., $\vec{\Omega} \cdot \sigma = C$, the vortex flux is constant. Where σ is the cross-sectional area of the vortex tube perpendicular to $\vec{\Omega}$. This requires that the cross-section area can't equal zero, otherwise the vorticity $\vec{\Omega}$ tends to be infinite. This means that concentrated vortices cannot be generated or vanished, i.e., interrupted, in the flow field. It can only be a curved cylinder with a ring shape or with both ends starting at the liquid level, or starting at the wall surface and or a cylindrical surface ending at infinity. Of course, there may be spherically concentrated vortices curved into rings. The generation of concentrated vortices can therefore be discussed in terms of their cross-section.

Focusing on the pressure trace on the wall (two-dimensional plane) under the no-slip condition, a good clue to the generation of concentrated vortices can be found.

Let x and y be the coordinates of the pressure values on the plane R^2. And the radial length in the plane from the origin should be the magnitude of the pressure. There is therefore the following non-linear self-consistent equation describing the movement trajectory of the pressure

$$\begin{aligned} \frac{dx}{dt} &= P\left(x, y\right) \\ \frac{dy}{dt} &= Q\left(x, y)\right) \end{aligned} \tag{3.4}$$

I.e., the qualitative analysis of such a system of nonlinear ordinary differential equations should find the condition that the pressure field needs to be satisfied for generating concentrated vortices. Thus $\frac{dx}{dt} = u \quad \frac{dy}{dt} = v$, note that here u and v is considered as the speed at which the pressure moves over R^2, and the resultant direction of these two velocities is the direction of pressure movement.

From spherical surfaces (ρ, θ) coordinates can be mapped to the R^2 plane with Poincare mapping

$$W = \Phi(z) = u(x, y) + iv(x, y) \tag{3.5}$$

This formula is meant following results of discussion to refer not only to a flat but also to any R^2 surface,

The following equations describe family of periodic solutions

$$\begin{aligned} \frac{dx}{dt} &= -y - x\left(x^2 + y^2 - 1\right), \\ \frac{dy}{dt} &= x - y\left(x^2 + y^2 - 1\right), \end{aligned} \tag{3.6}$$

$$y^2 + \omega^2 x^2 = C^2 \tag{3.7}$$

Coordinate transformation is written as

$$\begin{aligned} \frac{d\rho}{dt} &= -\rho \qquad \left(\rho^2 = 1\right), \\ \frac{d\theta}{dt} &= 1, \end{aligned} \tag{3.8}$$

There are two solutions $\rho = 0, \rho = 1$ to this set of equations. The origin is a singularity, the property depends on the characteristics of its linear part. This may be a critical node, corresponding to the linear part of the equation

$$\frac{dx}{dt} = x$$
$$\frac{dy}{dt} = y \tag{3.9}$$

Or one may be a central singularity, corresponding

$$\frac{dx}{dt} = -y$$
$$\frac{dy}{dt} = x \tag{3.10}$$

The concentrated vortex formation process cannot first produce a periodic solution such as the singularity like the center point. Thus, for pressure, it should be a critical node that results physically in the negative pressure field. That is, there is a singular point with negative pressure, and vorticity may be gathered around the point. Along with the collection of the vorticities, the reconstruction process of the spatial distribution of the translational momentum and the rotational momentum of the particles in the flow is also the evolution process of the pressure field, and the change generated by the nonlinear part corresponds to the periodic solution characteristics, so that the critical nodes can also be converted into center singularities.

In the ring region between the central singularities, periodic solutions such as limit cycle may occur over the ring region between the two closed-loop lines, taking into account possible axial flows in concentrated vortices. This periodic solution may be stable. It can be determined that the pressure field formed by the concentrated vortices must be present in the flow field as a $\Delta\rho = 0 - 1 < 0$ negative pressure zone.

Such a progression from a stationary point (critical node) to a limit cycle can be observed during the generation of Taylor vortex cells between two rotating cylinders drums. The Taylor vortex cells were observed with a laser source with a very slow start-up procedure, as observed the cross section of Taylor vortex cell during the formation of the first pair of vortex cells on the cylinder's wall above the rotating floor. The topology of the flow trace can be seen whether displayed with the tracer or measured with PIV [4].

3.1.4. The Sufficient and Necessary Conditions Generating the Concentrated Vortex

According to the principle of vorticity conservation under the barotropic fluid, conservative force and non-viscosity, the sufficient conditions for the concentrated vortex generation naturally require the prior vorticity field. Otherwise no concentrated vortices will be generated. In the boundary layer, viscosity is a main cause of generation of a vorticity field. However, the conservation of vorticity is a motion with the body, though vorticity field is more suitable for Lagrange description, but is also possible to take into account only the translational movement in the pressure field by used of the Euler coordinate. The particles may be either rotating or non-rotating. Thus, the conservation of vorticity can be ensured.

Sufficient conditions for concentrated vortex generation are: there is a set of particles with vorticity in the flow field.

Particle vorticity is defined in terms of the circular quantity, i.e., rotational motion in a two-dimensional plane about the center of mass. Think about $p\not{V}_{Mac} = TS_{B,\omega,\, Mac}$ this symmetry, i.e., the invariance of the system interactions, therefore the force arising gather process of rotating particles can only be pressure. A two-dimensional negative pressure gradient area in the translational momentum field generates flow to assemble.

The necessary conditions for generating the vorticity gather are: two-dimensional negative pressure gradient areas appear in the rotation flow.

There are spin particles and non-spin particles which are not continuously distributed in the fluid field when the vorticity togather in negative pressure area. This explains why there is difference two regions with vorticity fluid and non-vorticity fluid in a concentrated vortex. As that we can see in the experimental observation and in the analytical solution of some concentrated vortex. Such as the Ranking vortex

$$\begin{aligned} \omega_\theta &= \frac{\Gamma}{\pi r_0^2} = \frac{2V_{\theta,r_0}}{r_0} && r < r_0 \\ \omega_\theta &= 0 && r > r_0 \end{aligned} \tag{3.11}$$

Wherein $\Gamma = 2\pi r_0 V_{\theta,r_0}$ is the circulation in region with vorticity, in which a Boltzmann entropy reduction process by the formula (3.2) undergone. The Ranking vortex has a solid vortex core area, and all particle vorticity is a constant value for $r < r_0$. But in the region of $r > r_0$, the rotation motion is a non-vorticity potential flow region.

The non-vorticity potential flow particles in the concentrated vortex also revolve around a common axis, but the no spin is still non-vorticity potential flow. The concentrated vortex appears discontinuity of macroscopic vorticity in a continuous translational velocity field. This has always been an inexplicable question in the analytical solution of concentrated vortex. But as described above, this discontinuity is a result of the conservation of vorticity being maintained. The particle rotational momentum governed by the shear stress remains a constant value in the solid vortex core, no new vorticity is generated, and conservation between forward spin momentum and revolution momentum after generation of concentrated vortex is maintained. The non-vorticity particles are entrained when the concentrated vortex is generated and are also kept conservative of vorticity and collected in the potential region of vortex.

In the process of momentum space configuration reconstruction there is an action of $\left(T\Delta S_B\right)_{M,\omega}$, particle spin momentum is converted into rotation translation momentum. And corresponding rotation kinetic energy $T_{M,\vec{\Omega}}$ around a common shaft like the co-coordinated rotation has a tremendous destructive energy due to greatly increases the circumferential translation velocity V_θ.

There should be an additional acceleration term in the N-S equation to collect vortices that may not be integral, namely

$$\iiint_{\forall, A\in\frac{\partial p}{\partial x}<0} k_{\vec{v}} \cdot (\vec{V}_i \times \vec{\omega}_i) d\forall_i \tag{3.12}$$

Since this spatial integral is an undefined volume integral over an undefined negative pressure gradient region

This spatial integral may be obtained by integral transforming between the time sequence and spatial domains. As in the large vortex simulation numerical method, iteration process of removal shearing stress is equivalent to a time series integral. Due to the fact that may result in concentrated vortices along the $k_{\vec{V}}$ direction of the flow trace.

When the variables discussed above for pressure are considered as particle coordinates, the trajectories are the corresponding particle streamlines and are easily discussed in connection with axial flow. Because the streamlines are of a stable topology, the pressure distribution is likewise stable.

The results of qualitative analysis using nonlinear differential equations looking at (3.9) and (3.10) on the polar coordinate plane are singularities of velocities on the streamlines. Unlike in the neighborhood of the origin of the negative pressure gradient where the pressure distribution goes from the limit circle to the origin, the critical node of the streamline is transformed into a central singularity that appears a ring domain between the concentric circles. Considering the stability conditions of this concentrated vortices on this basis the relationship between the vorticity discontinuity in the radial direction and the axial flow is found.

3.1.5. Stable Limit Cycle

According to the concentrated vortex theory solution and the distribution of the actually measured pressure, while $t \to \infty$ the limit ω set and $t \to -\infty$ the limit α set of the trajectory are both on the limit cycle. If the origin is a stable focus point, the trajectory needs to flow to the origin. This must be determined according to the direction of the velocity $\vec{V}_r$. Therefore, the axial flow flowing in at the origin and the axial flow flowing out from the inner and outer side of the solid vortex core can ensure that the limit cycle is stable. Then the focus point is unstable, as a stagnation point and zero momentum of the vortex axis, it is likely that this is the primary cause of concentrated vortices having unsteady vibration momentum. And the stable limit cycle is the mapping of the Sullivan vortex in the R^2 plane due to the double cells structure formed by the axial flow.

The Taylor vortex cells between the rotating cylinders can also see a complex limit cycle topology on the cross section, and the flow trace topology can be explained only by the axial circulation of the vortex center.

3.1.6. Rectification Effect by Spontaneous Degeneration in Concentrated Vortex

The formation of concentrated vortexes degenerates the mass point vorticity and is a Boltzmann entropy reduction spontaneous process. Each particle vortex generated by the wall shear is a spin microstructure. The vortexes of mass point are spontaneously combined into a circulation flow field with a solid vortex core around a common rotation center under the action of negative pressure gradient. The momentum space configuration potential energy of the corresponding particle vorticity can be changed into stronger vortex circumferential translation velocity, namely have

$$T_{\omega,M}\Delta S_{B,\omega,M} = T_{\omega,M}k'\ln\left(1/N_M\right), << 0 \tag{3.13}$$

$$S_{B,\omega,M}\Delta T_{\omega,M} << 0 \tag{3.14}$$

$$S_{B,\omega,M}\Delta T_{\omega,M} = -S_{B,v,M}\Delta T_{v,M} >> 0 \tag{3.15}$$

Wherein b is the number of molecules defining a mass point or ensemble and has $k' = k_B / b$; $S_{B,v,M}$ is the number of fluid mass points with translation momentum in the concentrated vortex, and $S_{B,\omega,M}$ is the number of fluid mass points with vorticity. Here $T_{\omega,M}$ is the kinetic energy of the particle vorticity and $T_{v,M}$ is the translational energy along circular flow in the concentrated vortex. The kinetic energy of the spinning vorticity freezes, corresponding to the release of such kinetic energy. Then the translation kinetic energy converted into revolution is greatly increased. These relationships represent the reverse cascade process from the trapped particles vorticity forming translational circulation flow with axial flow, and the origin of the increasing circulation flow velocity.

Resulting in an increase of centrifugal force, it produces a $\Delta p > 0$ pointing out of the vortex. This is in opposition to the negative pressure gradient that forms the concentrated vortices. There will be a periodic pressure build-up in the direction of the radius of the concentrated vortices.

This periodic pressure variation first creates a further rectifying effect on the surrounding flow field. The periodic pressure is a pressure vibration source and can generate a mass point wave motion in a surrounding flow field, and wave motion of the mass point conveys the mass point vorticity to the negative pressure center of concentrated vortex. Since the distance space between the particles are sufficiently small during the compression cycle of pressure that as the vorticity transport between particles can yet occurs, and during the relaxation cycle, no such exchange of vorticity between particles occurs. This constitutes a rectifying effect of the vorticity. The vorticity quantity is continuously transported by the wave to the negative pressure center of the concentrated vortex. Like as after typhoon and tornado are generated, stronger storms can be continuously developed and generated.

This periodic pressure has a rectifying effect on the molecular rotational momentum within every mass point in the concentrated vortex. As previously mentioned, the momentum of molecular rotation corresponds to the internal pressure in van der Waals forces. When the extrusion action to the central vortex center is large enough, the internal pressure can act to transfer the rotation momentum among molecules; conversely, when the pressure is directed out of the vortex, the internal pressure does not transfer molecular rotational momentum. This constitutes a rectification of the rotational momentum between the molecules under action of periodic internal pressure by means the pressure in whole flow field. The microscopic random states on the molecular level should be degenerated as one state about the vorticity of this particle. This periodic pressure would constitute a complete reverse cascade process chain.

The particle compression period, in which the molecular rotational momentum spontaneously merges through the process of $\Delta S_{B,\omega,m} = k_B \ln(1/b) << 0$, greatly enhances the particle vorticity and produces the effect of heat release. But in the low-pressure period, entropy increasing process of $S_{B,\omega,M} \to S_{B,\omega,m}$ can occur in particle. The process should correspond to the process of absorbing heat $S_{B,\omega,m}\Delta T_{\omega,m} >> 0$ from the molecule. This thermal effect is similar to phase change. There is a thermal effect in the concentrated vortex generation and development process.

Finally, the periodic pressure on the particles is also the vibration force between the particles. The Hook force along the molecular vibration motion is rectified to the periodic pressure, and arise a sound wave with a certain frequency corresponding periodic pressure between particles. It is the whistling sound that can be heard in the concentrated vortices. But also, the exothermic and endothermic effects of the negative pressure cycle when the molecular vibration momentum degeneracy occur at different pressure cycles. This plays an important role in the vortex tube refrigeration effect.

Of course, this periodic pressure also directly causes the axial flow to be non-steady flow. Thereby periodic pressure could also cause a pressure pulsation in the axial direction. The limit cycle topology should therefore be an unstable manifold.

3.2. VORTEX TUBE REFRIGERATION AND R-H EFFECT

The air inlet is arranged at the asymmetric middle position of the long and straight pipeline with air flow outlet at either end. And high-speed airflow is pressed along the tangential direction. One outlet of the end is located in the center of the pipe, cold gas flow out, and the annular outlet of the other end flows out of the high temperature gas flow along the pipe wall respectively. The total temperature separation effect generated by the rotary flow is called as R-H effect. The baffle at the center of the outlet end of the hot gas flow will turn the part of the gas flow to 180^0. As it flows out from the other end the temperature is reduced to a cold gas flow. This vortex flow has two characteristics: 1. There is a turbulent boundary layer on the wall, and the center of the tube is a concentrated vortex similar to Ranking vortex but with axial flow; 2. Sharp whistle appears.

This flow has problems in common with the thermo-acoustic effect produced by heating the middle of a metal tube with two open ends as well as with the refrigeration of the pulse tube, which are difficult to interpret with existing knowledge of hydrodynamics and thermodynamics. In this case, the interaction invariance relationship $\Delta\left(p\not{V}\right)_{Mac} = \Delta\left(TS_B\right)_{Mac}$ is used to attempt finding the relationship between thermal effects when the macroscopic momentum is rebalanced with interaction within the continuous medium

3.2.1. Thermo-Acoustic Effect

A sound emitted in the tube can be heard when an asymmetric position in the middle of the long straight round tube is heated up. The frequency of the sound waves audible to a human should be within 20-20000 Hz. This indicates that heat generates the vibration momentum of air particles rather than a flow corresponding translational momentum. The heat raises the local pressure in the tube to produce translational momentum to exhaust the gas in the tube to the two ends of outlet, but the distance between the heating position and the two outlet ends is asymmetric to produce pressure difference at the two ends of the long tube to suck the gas at the one end. The real flow can be offset, and the pressure of one end in the pipe becomes unsteady and results in a gas periodic pressure that is initial source of vibration. Under the action of a temperature field, when ΔT appears in a macroscopic temperature field and $\Delta\left(TS_B\right)_{Mac}$ potential energy reconstructs a three-dimensional random vibration momentum field of mass points or ensembles, the states with weak vibration momentum can be degenerated by the states with strong vibration in a high-temperature region and can be done into single-macroscopic state of periodic translation momentum. This process releases heat because it is a spontaneous process of Boltzmann entropy reduction. The local heat flux will flow back to the high temperature area to generate reverse and arise inverted $\Delta T_{rev} < 0$ and $\Delta p_{rev} < 0$. This is the same process as wave propagation of heat conduction in the crystal. The heat wave is step by step propulsion process degenerating ensembles random translation momentum and constructs a periodic pressure field. The periodic force acting on the fluid mass point generates rectification effect on the translation momentum and the vibration momentum. I.e., states for N_M mass points with particle translational and vibration momentum in long tube degenerate as a macro state with overall periodic translational momentum and have $\Delta S_B = k' \ln 1/2N_M << 0 \quad \left(k' = k_B/b\right)$. This release of mass point translational and vibration momentum creates a backpressure within the tube that is opposite to the pressure differential within the tube created by the heating zone and the end backpressure differential, thereby creating an overall periodic translational force within the thermo-acoustic tube.

And the periodic force acting on the mass point has rectification effect on the translation and vibration momentum of the molecules in the mass point. That is, as the number density is increased until the internal pressure starts to work, the kinetic energy of b molecules in an ensemble or a mass point can be frozen. Become the mass point translation kinetic energy. Heat is released via $\Delta S_{B,,v,m} = k_B \ln(1/b) << 0$. When the particles pass through the pressure relaxation phase, heat is absorbed from the molecules in the particles, corresponding to $\Delta S_{B,v,m} = k_B \ln(b/1) >> 0$. This causes the particle to become an internal heat source term that may have like $\sin T$. It is primary reason amplifying the thermal effects of ensemble and creasing heat propagation by form of dispersion wave. The molecules vibration is propagated to generate sound wave.

The vibration momentum on three movement levels in the thermo-acoustic effect is central to the problem. This pressure does not generate translational momentum and therefore this problem is not a flow problem. Whistle indicates appearance of particle vibration of continuous medium. Thermal effects are the result of the degeneracy of microscopic states about vibration momentum within a particle and between particles.

3.2.2. Pulse Tube Refrigerator

When the energy conversion of the thermo-acoustic effect is reversed, the pulse tube refrigeration technology is generated. The whole translational momentum reciprocating in the pulse tube is created by the reciprocating motion of the piston. Because of the buffer of the gas reservoir with large volume and the block of the heat accumulator for the flow, the reciprocating motion of the piston has very little gas compression ratio, so there is no adiabatic effect. The reciprocating motion of the piston generates only periodic translational flow that can be considered as whole periodic vibration motion. The state numbers of N_M mass points pushed by the piston with the random three-dimensional translational momentum change to one macroscopic state with periodic translational motion. While this process is mandatory, it is a Boltzmann entropy reduction process.

$$\Delta S_B = k'\ln 1/N_M << 0 \quad \left(k' = k_B/b\right) \tag{3.16}$$

The energy amount released in the process arises a $\Delta T_{rev} < 0$, and by the principle of force invariance is adopted, so that there is the reverse action $\Delta p_{rev} < 0$ that is opposite to $\Delta p > 0$ that points to the direction of the piston motion, and the reverse action maintains relation $\Delta p_{rev} >> \Delta p$. The periodic action that produces the rectifier effect is therefore not an action on the piston. There is reverse process when the reverse motion of the piston occurs. And a strong periodic force Δp_{rev} is induced in the whole periodic translational fluid forced by the vessel. The Δp_{rev} on both sides of the piston are balanced and no additional resistance is generated to the piston motion. This periodic force produces periodic translational motion. It is this force that also produces a wave transport of heat in the pulse tube. And the heat source item of temperature results from the periodic pressure acting on the fluid mass points. During the compressing period of particle and the number density reaching a certain level the molecular translational momentum in a mass point or an ensemble will be frozen and degenerated into a particle translational momentum state. And the internal pressure will play a main role. There is

$$\Delta S_{B,\Delta p_{re}<0} = k_B \ln\left(b/1\right) << 0 \tag{3.17}$$

The energy released is sufficient to raise the temperature of the particle. However, in the negative Δp_{re} stage, $\Delta S_{B,\Delta p_{re}<0} = k_B \ln\left(b/1\right) >> 0$ is enough to generate strong effect of absorbing heat from surrounding particles and medium to reduce temperature. A $\sin T$-like heat source term may thus be present in the gas particle of pulse tube

So that the gas particle thermal wave equation similar to the crystal heat conduction wave equation can be redefined as the thermal wave equation of a thermal vibration source with the length consisting of the n' mass points

$$\frac{\partial^2 T}{\partial t^2} = c_{0,T}^2 \frac{\partial^2 T}{\partial x_i^2} \tag{3.18}$$

Here $c_{0,T} = \omega_0 l_0$, $\omega_0 = \frac{\beta}{\alpha} = \frac{n'k}{\rho c_v}$ are respectively the wave speed and the frequency of the thermal wave. It can be seen that longer vessels have a larger number n' of mass points and that the length of the source of thermal vibration can produce lower cooling temperatures associated with larger thermal wave velocities. And the physical properties of the gas: the thermal conductivity k, density ρ and specific volume c_v all have an influence on the wave velocity of the thermally conductive wave, and thus the refrigeration effect.

The small holes can produce phase-adjusting effect on the two fluctuation processes. The pressure phase in the two sides of the piston appears alternately, and the bypass can help all to release the heat of two side pressure process into the high temperature heat exchanger, and to recovers the cold amount in low temperature heat exchanger.

When the periodic temperature vibration source for retention gas mass point in the plus tube is written as a sinusoidal term of temperature, so the corresponding wave equation should be sine-Garden equation

$$c_{0,T}^2 \frac{\partial^2 T}{\partial x^2} - \frac{\partial^2 T}{\partial t^2} = m^2 \sin T \tag{3.19}$$

This is a wave equation with a solitary wave solution as follows. [15]

$$
\begin{aligned}
&T = 4\tan^{-1}\exp[m\gamma\left(x - vt\right) + \delta] \\
&\gamma^2 = \left(1 - v^2\right)^{-1} \\
&T_x = 2m\gamma\ \sec h[m\gamma\left(x - vt\right) + \delta] \\
&\sin\frac{T}{2} = \sec h[m\gamma\left(x - vt\right) + \delta]
\end{aligned}
\tag{3.20}
$$

The kink solution of the positive γ value and the anti-kink solution of the negative γ value do not annihilate and do not change their shapes at the time of collision. Moving in the opposite direction, this accords the solitary wave solution characteristic of the K-dV equation. The linear waves propagate only an infinitely small amount of energy, but the solitary wave has the ability to propagate limited energy without energy losing. Therefore, it is the decisive reason for the deep separation of the total temperature with limited amount of pressure and temperature different. This is further demonstrated in the following discussion of the R-H effect generating total temperature separation refrigeration in the vortex tubes. Since the mechanical work produced by the reciprocating movement of the compressor piston is misunderstood to contribute the thermal effect when the gas expansion and compression, it is explained as a key cause of pulse tube refrigeration. In practice only a pulsating whole translational movement is provided which is necessary for thermal wave propagation.

3.2.3. Vortex Tube Refrigerator

The vortex tube has a gas inlet to create a rotational flow and a hot and cold two gas outlet. There is no mechanical reciprocation. When sound waves appear in the vortex tube, the refrigeration effect is achieved. Therefore, need to search for how the sound waves are generated in the vortex tube?

The gas entering the round tube tangentially is a non-vorticity fluid that develops the point vortex flow. It is still non-vorticity. The outflow is the whole macroscopic translational momentum of the gas. Under the condition of no slip boundary on the wall surface, the particle shear stress generates the particle spin momentum that is vorticity of mass point. Through Boltzmann spontaneous reduction process all vorticities degenerate into a concentrated vortex. And constitutes a radial periodic pressure differential. This periodic pressure creates a rectifying effect at the level of the whole translational motion level as well as at the level of particle motion and molecular motion. But also degenerates the translational, rotational and vibration momentum. See section 3.1 for a discussion about concentrated vortexes. The mechanism by which this periodic pressure difference produces a thermal effect in the vortex tube is discussed with emphasis here.

The periodic force on the particles corresponds to the Hook force on the particles that generated the acoustic wave. The device has rectification effects of random three dimension molecular vibration momentums and spin momentum within the particle, i.e., as the pressure is increased to the internal pressure the molecular momentum within a mass point is frozen and then changed into spin momentum and vibration momentum of the particle. During the pressure release period, a Boltzmann entropy-increasing process occurs and the particle momentum returns to the molecular momentum state. So that the process of releasing heat and absorbing heat alternately occurs. Not only do vibrations of the particle to emit sound waves, the particle also becomes an internal heat source item of the $\sin T$ type. The particle also becomes a $\sin T$ type internal heat source term.

$$\frac{\partial^2 T}{\partial t^2} - c_{0,T}^2 \frac{\partial^2 T}{\partial r} - \frac{1}{r}\frac{\partial T}{\partial r} = m^2 \sin T \tag{3.21}$$

The circumferential velocity in the concentrated vortex is orthogonal to the temperature gradient, and the radial velocity can be ignored. The heat convective term can be omitted from the (3.21) equation. And it can be written as a hyperbolic wave equation of heat conductivity. There should be a dispersion wave solution. Low temperature gas flow appears at the center region of vortex, and high temperature gas flow does at the tube wall region.

The total temperature separation in the gas flow should be a thermal propagation process by wave form with limited temperature vibration amplitude. Practical cases support solitary wave transport with limited temperature different. There will eventually be a finite density difference at the center and the tube wall.

The baffle of heat flux returns the axial flow of vortex core to form a Sullivan vortex having double cells structure. Its manifold on the R^2 plane is an unstable limit cycle. These instabilities of the vortex flow form pressure pulsations in the radial direction. This instability will also have a coupling effect with the axial with the axial pressure pulsation caused by pressure asymmetry at the two ends of the outflow. It is easier to generate sinusoidal thermal vibration source terms in the gas staying in the tube under the condition of outflow (no outflow in pulse tube refrigeration, and cold and heat output by the heat exchanger), similar to in the pulse tube. The gas remaining in the vortex tube would constitute a temperature vibration term $m^2 \sin T$ similar to that in (3.19) equation. And it is also sine-Gordon equation about heat wave transfer.

Vortex tubes refrigeration mechanisms differ from pulse tube in that: the pulse tube has no net translational flow, only makes periodic whole translational motion, rectifies and responds to molecular translational momentum and vibration momentum, and generates solitary wave transportation of heat. The solitary wave in the vortex tube generates periodic pressure after forming concentrated vortex accompanied by a continuous translational flow field, and a sinusoidal heat source is generated due to the rectification effect on the rotation and vibration momentum of molecules. The thermal effects of mechanical compression and expansion in pulse tube refrigeration are also negligible.

3.3. Energy Source of Destruction Generated by Resonance of Solid Structures

Solid structural, particularly for the bridges and rotors of turbo-machine, are subject to structural failure due to resonance. It is also common knowledge of engineers to resonate at an excitation frequency equal to or very close to the natural frequency of the structure. It is unclear where the tremendous energy originates, however, in what way the resonance generates destructive forces. This section attempts to discuss this problem from the potential energy $\Delta\left(TS_B\right)_M$ that the elastic forces of the particles in the solid structure have.

The elastic force of the mass point of the structural component can enable the structural component to generate vibration in the elastic deformation range. Such vibrations generate propagation of elastic waves for the structural element. Such as vibrations in the structural elements at various locations of the bridge are propagating in their respective frequency, phase and direction. The vibrations generated by the different external force on the bridge are randomly superposed everywhere. The random vibration also continuously reconstructs the vibration momentum on the whole bridge over whole space, and certain dynamic balance is restored. The elastic force of the mass point in the structural member is the interaction for rebalances the momentum distribution in the structural elements by the space configuration potential energy $\left(TS_B\right)_{L,\,Mac}$ to correspond the vibration momentum of the mass point. From the discussion of the resonance phenomenon, it can be seen that this potential energy does exist.

When the excitation frequency is close to or the same as the natural frequency of the structural element, the displacement peak value and the speed peak value of the forced vibration reach limited values according to a forced vibration equation. This kinematic and dynamic result does not imply the necessary damage of the structural elements. There must be condition for centralizing all discrete maximum vibration kinetic energies in order to form the very large energy that can destroy the structure. However, there is no such equation or condition related to the accumulated resonance energy in classical mechanics.

3.3.1. Spontaneous Reverse Degeneracy of Boltzmann Entropy Reduced under External Force

A team of soldiers is walking through the bridge and still has a negligible external force in terms of weight. Moreover, no upward reverse force exists, and finally no vibration force and resonance are generated for the bridge deck. Without such a vibration source, resonance cannot propagate through the wave and the entire bridge.

Suppose that N' structural units on a bridge are subjected to the periodic action of external force, and the vibration momentum of N' different units will change into 1 vibration momentum with the same size and direction under the periodic external force. All solid ensembles or mass points in every unit have N_{ens} numbers of particles, and degeneracy of the original random momentum of these particles as the macroscopic vibration momentum of the same frequency. The process is a spontaneous process of Boltzmann entropy reduction. In the initial stage there is for N' structural units $\Delta S_{B,ens} = (N'-1)k_B' \ln(1/W_{ens}) << 0$. This change corresponds directly to the release of particle vibration kinetic energy $T_{0,ens}\Delta S_{B,ens} << 0$. The force invariance should be given by

$$T_{0,vab.ens}\Delta S_{B,ens} = \not{V}_{ens}\Delta p_{ens} = \not{V}_N \Delta p_{Sys} << 0 \qquad (N_{ens} \in N) \quad (3.22)$$

I.e., very large opposing forces $-p_{ens}$ and $-p_{Sys}$ are generated between the ensemble points and between the building structure units. Under the resonance condition the reverse acting force produce rectification effect that enables the N' structural units to become resonance vibration sources. This vibration is propagated in wave form throughout whole structure. Meanwhile, the vibration acting force between the structural units and mass points generates a spontaneous degeneration effect of the vibration momentum between the structural units and the ensemble in the units and the intermolecular in the ensemble. These are all Boltzmann entropy decreasing processes on three motion levels. For the N structural unit, and the N_{ens} ensembles in each unit, as well as W molecules in each ensemble all undergo state degeneracy in resonance occurring under this rectification force, the Boltzmann entropy of the last order should be as $\Delta S_{B,vab} = k_B \ln(1/W)$. And the vibration energy finally all release out. This is the energy that causes the destruction of the entire system structure.

If each element in the structure has a stiffness coefficient, it may be considered a macro state and the entire structure may be considered a statistical system having $N_{L,\,Mac}$ macro states. The conditions for resonance are: for a sufficiently large number $N_{L,\,Mac}$ of macro-states, when the respective forced vibration frequency of each structural element changes is changed into the same forced vibration frequency equal to natural vibration frequency, the macro vibration state become to only one vibration state of all macro-structure units. The process can be regarded as a Boltzmann entropy spontaneous reduction process, and for macroscopic vibration kinetic energy, the process should have

$$\Delta S_{B,\,L,\,Mac} = (1 - N) \tag{3.23}$$

And the vibration kinetic energy of the maximum peak velocity at resonance should be

$$T\Delta S_{B,\,L,\,Mac} = \tfrac{1}{2}(F/\gamma)^2\, N(1 - N) \tag{3.24}$$

Where F is the mode of the periodic force and γ is the linear resistance coefficient of the vibration velocity.

In addition, the vibration kinetic energy released when the microscopic states of the ensemble or the mass point are merged and the molecular microscopic states are degenerated in the ensemble

$$T_{ens}\Delta S_{B,ens} = NT_{ens}k'_B \ln\frac{1}{N_{ens}} \qquad (for\ all\ N, \quad k' = k_B / W) \quad (3.25)$$

$$T_{Vab,moll}\Delta S_{B,Vab,mol} = N_{ens}T_{Vab,mol}k_B \ln\frac{1}{W} \ \left(W \in N_{ens}\right) \quad (3.26)$$

The resonance condition is the necessary condition for the vibration kinetic energy to gather and generate damage: the frequency of the forced action is equal to the natural frequency of the structural elements. Namely

$$\omega = \omega_0 = \sqrt{k/m} \quad (3.27)$$

Sufficient conditions when resonance occurs are: each element in a solid structure has the same or similar natural frequencies. This means that the elements of the structural member have the same mass and geometry and also need to be of the same material, i.e., have the same stiffness coefficient. The structural elements of the bridge and the impeller of the rotary machine naturally satisfy this condition, and therefore, resonance destruction is particularly likely to occur.

When the distribution of the vibration momentum in the space is reconstructed by the $\left(TS_B\right)_{Sys}$ potential energy, the maximum resonance energy is gathered at the position with the weakest structural strength, the maximum displacement is generated, and the damage arises.

The problem of the concentration of energy in resonance is discussed in the hope of accounting for the spatially distributed potential energy $\left(TS_B\right)_{Sys}$ of macroscopic elastic vibration momentum present in motion on a macroscopic level. There are also $\left(TS_B\right)_{ens}$ and $\left(TS_B\right)_{mol}$ at the ensemble and molecular level. The spontaneous reverse action generated by Boltzmann entropy reduction generated by the state numbers on each level under the action of external force releases the energy of the vibration momentum of each level and generates the energy for destroying the macroscopic structural member.

Chapter 4

TS_B POTENTIAL ENERGY OF QUANTUM STATE IN ATOMS AND MOLECULES

The preceding discussion refers to condensed states, microscopic and macroscopic motions involving a plurality of atoms or molecules. The chemical reaction process is discussed in physical chemistry. The most basic process in chemical reaction is the process in which electrons outside the nucleus between individual atoms produce the molecules reconstituting electron pairs based on Coulomb force. Or in turn, the process of splitting the electron pair into atoms. Here the interaction is an electromagnetic interaction in the standard model. The object of the problem is a single atom or molecule. This chapter mainly discusses the translation momentum of electron around the nucleus. Of course, the multi body problem of the chemical reaction is discussed in thermodynamics by the concept of chemical potential. Here we discuss whether there is a potential energy $\langle TS_B \rangle_e$ of the electron momentum moving around the nucleus under the action of Coulomb force.

4.1. Chemical Potential

The changes of composition in the system include physical mixing process and chemical reaction, both of which generate thermal effect. As far as heat is concerned, they can be placed in free energy. However, the change of components in the system represents the change of the system itself, so a chemical potential is needed to represent it. The partial molecule Gibbs free energy for component A that is chemical potential μ_A can be defined as

$$\mu_A = G_A = \left(\frac{\partial G}{\partial n_a} \right)_{T,p,n_A} \tag{4.1}$$

$$dG = Vdp - SdT + \sum_A \mu_A dn_A \tag{4.2}$$

The motion momentum at different levels in the above definition is put together. There may be different momentum participation in the thermal effect, and there may be at least two kinds of momentum taking part in the change of the number of particles in the system. So that in addition to convenient calculation, it will cause confusion in understanding. The heat effect has both mixing and chemical reaction, for example, can generate heat effects that also change the particle number of components in the system. But the two processes are governed by different interaction. The same interaction is the premise of energy and momentum conservation. So free energy can't be conserved. All thermal effects can't be summarized by a first law of thermodynamics in the form of heat-work equivalent. Since the principle of minimizing the free energy is the most important basis of research of life science, especially biophysics. Because free energy itself has a lot of uncertainty, it is not surprising that there are difficulties in application. It is the important reason for discussing the topic.

The chemical reaction is the process of electron pairing in outer orbit, and relates to the Coulomb force and electromagnetic interaction. Schrodinger equation for discussing various momentum of electron motion would be a good starting point of problems.

4.1.1. Statistics for Space in Electron Cloud

The momentum of an electron in orbit is determined by the discrete quantum numbers by solving the Schrodinger equation. Only the principal quantum that determine the orbital shape and distance from the nucleus are discussed without loss of generality of the problem. The electron orbit of classical mechanics is an electron cloud determined by probability space in quantum mechanics. But electrons in the electron cloud are continuous for time. The statistics of the spatial positions make the momentum field of the electron moving around the nucleus by the Coulomb force between atomic nucleus and the outer electron, that the cloud is a distance space $\not{V}$ with statistical significance. The Coulomb force between charges should be the interaction p in the spatial position configuration potential $\langle p\not{V}\rangle_e$ of the statistical system. The subscript e refers to the system for electron.

The electron cloud spatial shapes for different principal quantum numbers are different, which means that the spatial position configurations for different momentums are different. It should also correspond to a momentum space configuration potential $\langle TS_B\rangle_e$. The electron motion also generates a magnetic field according to Maxwell equations. There is thus an electromagnetic field around the electron cloud, which should also be the field energy $\langle TS_B\rangle_{em}$ in the electromagnetic field generated by the electron motion.

The statistics in distance space of the rotation of a single electron around a nucleus are equivalent to the statistics for distance space about set of molecules. The equivalence of two kinds of statistics also shows that the statistics about state number all have determinacy. The macroscopic variables under the thermodynamic equilibrium state in the condensed matter are determinacy. And the electromagnetic spectrum of the electrons at the transition between different electron clouds is certainty. And the property will not change due to random spatial locations. Therefore, the classical expression of thermodynamics for electrons motion in atoms or molecules is still used owing to such quantum systems are also a deterministic statistical system.

The electron motion around the nucleus is a revolution, called the isotopic spin in the standard model, to distinguish the spin of the electron. The isotopic spin conservation of electron corresponds to the gauge invariance, which means that the kinetic equation remains unchanged when the phase factor is used for conversion which means conservation of angular momentum.

The Coulomb force between the electron and the nucleus is a conservative force. The conservation of mechanical energy is satisfied, i.e., the sum of the kinetic energy of the electron and the position configuration potential energy between the electron and the nucleus is a constant. There is little doubt as to the existence of such $\langle p\not{V} \rangle_e$ potential energy. Born lays an important foundation stone of quantum mechanics about the kinetic energy in a hydrogen atom model by quantized discrete values.

Is there also a momentum space configuration $\langle TS_B \rangle_e$ for the electron cloud around the nucleus? In terms of Maxwell's thermodynamic relationships $\langle p\not{V} \rangle_e = \langle TS_B \rangle_e$ and the resulting in invariance of interactions, there should also be potential energy $\langle TS_B \rangle_e$ under Coulomb force. As long as there is the position space potential energy $p\not{V}$ under some force, and has corresponding TS_B potential energy.

$\langle p\not{V} \rangle_e$ Electromagnetic spectra of hydrogen atoms and molecules are discussed to confirm that the probability of spatial distribution of momentum in the electron cloud corresponds to the $\langle TS_B \rangle_e$ potential energy of electron movement around the nucleus.

4.2. $\langle TS_B \rangle_e$ POTENTIAL ENERGY IN H ATOM

The discovery of the electromagnetic spectrum of hydrogen atoms is of great significance in driving the development of modern physics. In order to maintain conservation of energy, it is necessary to be able to tell where the energy of such electromagnetic radiation originates. It can be assumed that the basic interaction between the electron in the hydrogen atom and the proton in the nucleus is Coulomb force. This is a conservative force. At the same time, the electron motion forms an electromagnetic field given by Maxwell's electromagnetic equations.

The Bohr's hydrogen atom theory confirms the concept of electron orbital separation, and the energy of the hydrogen atom electromagnetic spectrum corresponds to the orbital energy difference calculated by quantization. A sharp problem thus occurs. After the electrons radiate energy, the electrons lose their kinetic energy and should collapse into nuclei, i.e., conservation of energy is trapped in the crisis. Bohr considers that photons are not radiated by electrons. Schrodinger interprets the spatial trajectory of electrons as an uncertain probability space by wave function, but the uncertainty of quantum mechanics cannot answer the energy conservation problem involved with photons.

Heisenberg proposed the concept of "commutant" for the purpose of explaining the strong interactions between protons. The popularization of Yukawa Hideki to mesons, including electromagnetic interactions, proves that the "commutant" of the electromagnetic field is a photon. That is, the energy of the "commutant" is equal to the energy of some interaction, which is equivalent to the force field to add another energy. This is equivalent to adding energy beyond that mechanical energy conserves by the original conservative forces. But radiation does not always occur with respect to a particular photon, and electromagnetic interaction field is a continuous and persistent existence.

The "commutant" theory acknowledges that atomic electromagnetic (photon) energy originates from the exchange with other energy. This energy exchange results from the exchange of the electromagnetic interaction of electrons with the Coulomb force of the electrons. Or it can be called as coupling action. The “commutant” is a strong physical idea. The method plays an important role in the standard model. Therefore, there is still a significant point in the deep discussion of the electromagnetic radiation problem of hydrogen atoms.

4.2.1. $\langle p\not{V} \rangle_e$ And $\langle TS_B \rangle_e$ Potential Energies of Electron Cloud in H Atom

The H atom electron cloud obtained by the Schrodinger wave equation are probability spaces, and are possible spatial positions of electrons moving around a nucleus under the action of Coulomb force between the electrons and protons. The electronic cloud is more consistent with the concept of distance space $\not{V}_e$. Namely the interaction determines the distance, and the electron cloud should be considered as a distance space. The location of the electron in the electron cloud is uncertain, so that the electron in the atom system has a position configuration potential $\langle p\not{V} \rangle_e$. Since the different positions of the electron cloud correspond to different electron momentums, the electron cloud is also a momentum space configuration. According to the symmetry of invariant force and the electron cloud is also a momentum space configuration, so the electron cloud also has momentum space configuration potential energy $\langle TS_B \rangle_e$. For the main quantum number, the distance space scale of the electron cloud has a deterministic statistical mean value, which can be inferred from the certainty of the spectral line. Although it’s a quantized split value, both experimental values on earth and those obtained in outer space are the same at any time. This shows that momentum and energy in electron cloud are not only certainty, but also conservation. The question is, what energy does the energy jump in electron cloud directly correspond to? We also need to prove that the other energy of the corresponding photons should be the change of $\Delta\langle TS_B\rangle_e$ in the electron cloud during the electron jump.

The momentum space configuration of the corresponding electron cloud has only concept and no expression. Its potential energy has never been recognized: as this is an unrecognized energy. It should no longer be recognized as a kind of heat. Although it is a radiative heat, but rather hope it is an energy. The orbits with different momentum are different shape from the corresponding electron clouds, and although the electron orbits of the outer layer of the hydrogen atom are all spherical, the electron clouds of the outer layer electrons of other elements are not all spherical. This is not covered by the track concept. Schrodinger equation reveals a new space, the position space, or distance space, depending on the size of the interaction. This is again a momentum space. The H atom has only one outer electron. Importantly, in a distance space, the probability of each point in the distance space is different, so that the momentum is not distributed in the space in an equal probability; this is a spatial configuration of momentum. It has potential energy. Symmetry $\langle p\mathcal{V}\rangle_e = \langle TS_B\rangle_e$ according to the Maxwell thermodynamic relationship should be present. The momentum space configuration of this electron motion between electron and nuclear has a corresponding potential energy $\langle TS_B\rangle_e$.

It can be discussed by a thought experiment that this $\langle TS_B\rangle_e$ potential energy is the energy source of electromagnetic radiation (photons), which can be heat in the form of light radiation. Imagine an isolated system consisting of an absolute blackbody and one H atom, which is insulated from the outside environment. The blackbody temperature is slightly lower than the kinetic energy of electron in H atom. There is the condition that H atom can radiate photon to blackbody. There is a completely elastic collision between an atom and blackbody. The blackbody does not cause mechanical deformation as a result of collision. The blackbody is also big enough not to change the temperature due to absorption photons. The change of Gibbs free energy for this isolated system can be written as

$$dG = dG_e + dG_v = dT_e + d\langle p\not{V} \rangle_e + d\langle TS_B \rangle_e + dT_V +$$

$$+ d(pV)_V + d(TS\ )_V \leq 0 \tag{4.3}$$

Here the subscript V refers to the blackbody container. The blackbody only absorbs the electromagnetic radiation energy obtained by photons

$$d(TS)_V = Q = h\nu \tag{4.4}$$

The mechanical energy is conserved between the electron kinetic on orbit and the interaction potential of the Coulomb force. Thus having

$$dT_e + d\langle p\not{V} \rangle_e = 0 \tag{4.5}$$

And there is follow relation

$$d\langle TS_B \rangle_e + h\nu \leq 0 \tag{4.6}$$

Because the electron number in H atom keep invariant only on-orbit momentum corresponding to the main quantum number is taken into account, and micro-state number remains the same during jumping time, so that the above relation can be written as an equation

$$\langle S_B \Delta T \rangle_e = -h\nu \tag{4.7}$$

That is to say, the electromagnetic radiation energy of the photon originates from the potential energy difference when the momentum space configuration of the electron cloud changes. It is accompanied by the jump of electron from the electron cloud with high momentum to low momentum electron cloud. The replacement V by $\not{V}$ in the Gibbs free energy of an electron means the concept of replacing work with potential energy;

substitution of $S_{B,e}$ for S is the concept of substituting the spatial probability of momentum for heat. The concept of work and heat in the classical thermodynamics for the container is still used without loss of consistency, as the free energy concept allows the emergence of energy at different motion levels.

According to the principle of analytical mechanics, the spatial configuration in any dynamics corresponds to the corresponding potential energy. Electron clouds of different energy levels around atoms not only have a distance space variation, but also have different time scales of corresponding momentum or velocity in the corresponding different distance spaces. Therefore, the electronic cloud not only has discrete position space potential energy $\langle p\not{V}\rangle_e$, but also has discrete momentum space configuration potential energy $\langle TS_B\rangle_e$.

4.2.2. Virlia Theory

The total energy H_e of hydrogen-like atom obtained by Schrodinger equation has the following relationship with the statistical average interaction potential energy $\langle p\not{V}\rangle_e$ of Coulomb force.

$$H_e = \frac{1}{2}\langle p\not{V}\rangle_e \tag{4.8}$$

Here p is Coulomb force, and $\not{V}$ represents the distance space between the electron and the core, that is distance r. Obey to $H_e = \langle T\rangle_e + \langle p\not{V}\rangle_e$, the kinetic energy of electrons on orbit in the - hydrogen-like atoms can be proved to be

$$\langle T\rangle_e = H_e - \langle p\not{V}\rangle_e = -\frac{1}{2}\langle p\not{V}\rangle_e \tag{4.9}$$

Virlia theorem: for the system of which the potential energy obeys the ar^m law, the kinetic energy and the potential energy satisfy the relation

$$\langle T\rangle_e = \frac{1}{2}\langle V\rangle = -\frac{1}{2}H_e = -\frac{1}{2}\langle p\not{V}\rangle_e \tag{4.10}$$

The m is equal to -1 about potential energy of Coulomb force in the hydrogen-like atoms, that has the relationship ar^{-1}. Therefore, the formula (4.9) proves that the Virlia theorem in hydrogen-like atoms.

The total energy of the hydrogen atom system is constantly not zero. There is $H_e \to 0$ at the $r \to \infty$. That is, there is no kinetic and $\langle p\not{V}\rangle_e$ potential energy of the Coulomb force field before the electrons is trapped by the hydrogen nuclei. As the electron approaches the core, the $\langle p\not{V}\rangle_e$ becomes smaller(the absolute value becomes larger) and the $\langle T\rangle_e$ becomes larger, the centrifugal force generated by the electron velocity can balance the larger Coulomb force, and electron can only not fall into the nucleus. Total energy H_e cannot be a non-zero value.

The problem with this result is 1. Coulomb force is a conservative force, why is H_e not conserved at zero? 2. Why is H_e negative value? What is the kinetic energy in Virlia theorem? In the theorem the kinetic energy is equal to half of potential energy of the Coulomb force. Can the mechanical energy of the Coulomb force conserve? 3. H_e <0 indicates that the electron gains energy during capture by the nucleus into the electron cloud. What energy was obtained?

To clarify these doubts, two potential energies are needed to correct the Virial theorem.

$$
\begin{aligned}
&H_e = \langle T\rangle_e + \langle p\not{V}\rangle_e + \langle TS_B\rangle_e \\
&\langle T\rangle_e + \langle p\not{V}\rangle_e = 0 \\
&H_e = \langle TS_B\rangle_e \\
&\langle p\not{V}\rangle_e = \langle TS_B\rangle_e \\
&\langle T\rangle_e = \langle TS_B\rangle_e - \langle p\not{V}\rangle_e = \frac{1}{2}H_e \\
&\left|\langle p\not{V}\rangle_e\right| = \langle TS_B\rangle_e \\
&\langle T\rangle_e = \frac{1}{2}H_e
\end{aligned}
\tag{4.11}
$$

Coulomb force is conservative force, and mechanical energy conservation and momentum conservation can be maintained. The total energy is thus the electron momentum space configuration potential of the electron cloud. But it is also an energy having positive value and therefore the energy absorbed as it approaches the nucleus. Due to the symmetry of the acting force, the kinetic energy is half of the sum of the two potential energies. The total energy is thus still equal to 2 times the kinetic energy in the Virlia theorem. There is still a quantitative relationship between energies in Virlia theorem. It means that there are two kinds of potential energy in the system, namely the aforementioned $\langle p\not{V}\rangle_e$ potential energy and $\langle TS_B\rangle_e$ potential energy, and Virlia theorem can well be understood. The $\langle p\not{V}\rangle_e$ potential energy is a probability that electron under the action of Coulomb force will occur at some location in electron cloud. And $\langle TS_B\rangle_e$ is the probability at which the corresponding velocity occurs. Where the probability of occurrence is high, the weight of this momentum in the kinetic energy $\langle T\rangle_e$ is high. Otherwise the weight is small. The $\langle TS_B\rangle_e$ can be used for atomic outer electron clouds because it is a statistical system. And its statistical average is determined for discrete orbit radius, so its number of states is countable.

The number of outer electrons of H atom is 1. $S_{B,H,e}=1$. This is consistent with the "countable" of Boltzmann entropy. The countable of TS_B is an important commonality.

Each of these two potential energies may have a corresponding change for kinetic energy. The mechanical energy of Coulomb force is conserved. While $\Delta\langle T\rangle_e$ varies, there is $\Delta\langle p\not{V}\rangle_e$, $\langle TS_B\rangle_e$ is varied with their changes. Of course, it does not directly change the orbital kinetic energy of the electrons. When the number of microscopic states is not equal to one, $\Delta\langle T\rangle_e$ is not equal to $\Delta\langle TS_B\rangle_e$. It should also be equal to the kinetic energy and the half negative total energy according to the principle of force invariance. There is more energy entering the electron cloud orbits near the nucleus, meaning that the outer energy is absorbed as $\langle TS_B\rangle_e$ changes, only there should be $\Delta\langle TS_B\rangle_e>0$. And only the reconstruction capability of the potential energy can keep the total energy conservation. But this energy is not that of the Coulomb force field. Only the energy in the electromagnetic interaction derived from Coulomb force. When electrons move around the nucleus, the moving electrons generate electromagnetic field energy. (After the manuscript is complete, the electromagnetic field for single particle can be experimentally confirmed and Maxwell electromagnetic field equations can be applied to singer particle). This field energy can be noted as $\langle TS_B\rangle_{em}$. Here $\langle TS_B\rangle_e$ as well as $\langle TS_B\rangle_{em}$ potential energies are two different momentum field configuration potentials with different interaction on the same carrier. But both are equal each other. Thus, when electrons transition, $\Delta\langle TS_B\rangle_e$ corresponds to the electromagnetic radiation generated by amount of $\Delta\langle TS_B\rangle_{em}$. A photon is therefore called a "commutant" and refers not only to the exchange of two interactions (i.e., coupling action) but also to the exchanger between two potentials. The introduction of the $\langle TS_B\rangle_e$ potential energy concept makes the "commutant" concept more clarity or concrete. While $\langle TS_B \rangle_e$ and $\langle TS_B \rangle_{em}$ are introduced to maintain the overall energy field and mechanical conservation. The Virlia theorem is a quantitative reasoning for Schrodinger equation hydrogen atom solution. There will be a result of $H_e < 0$ in the solution of Schrodinger equation with Coulomb force only. Because the $H_e < 0$ condition is introduced in the determining solution conditions, the above discussion is therefore well founded. Since the determining solution conditions can also be set to $H_e > 0$ or $H_e = 0$.

Electromagnetic interaction plays an important role in recent physical development, and it is necessary to examine the above-mentioned findings. Since the thermal effects of chemical reactions and radiation are covered by Maxwell thermodynamic relationships. It is possible to consider these thermal effects as TS_B potentials corresponding to a certain momentum. In quantum mechanics, the coupling between different interaction forces is considered as a commutant. The connection between the two is not only necessary, but also beneficial to expand the application range.

The commutant is a boson with energy. The photon is a kind of boson. The difference between boson and fermion is that the spin quantum number is an integer and does not need to satisfy the Pauli incompatibility principle. Many bosons can be in the same quantum state. However, in the statistics of microscopic states about $\langle TS_B \rangle$, particles in the indistinguishable region of local space are allowed to have spatial degeneracy states. Therefore, $\langle TS_B \rangle$ should be applicable to both bosons and fermions. It's not limited to bosons. The commutant referred as the energy with particle property is appropriate because $\langle TS_B \rangle$ could not be of mass. They can be an energy field of the bosons or fermions. It can be energy field with mass or without mass. As a result, during $\Delta \langle TS_B \rangle$ conversation process there may show itself a particle without mass, but must be a particle having energy. The two concepts of commutant and $\langle TS_B \rangle$ energy field are compatible at least and may be the same.

When the energy quantum transition of the on-orbit electron occurs, Coulomb force is conservative force. And it is conservative of mechanical energy between potential energy and kinetic energy. The extra energy of commutant, therefore, is the previously emitted or absorbed electromagnetic field energy in the form of $\Delta\langle TS_B \rangle_{em}$, converted to energy in the form of photons by $\Delta\langle TS_B \rangle_e$ energy transitions. They have different energy fields. The photons appear as particles. The $\langle TS_B \rangle_{em}$, where electrons move around the core, is energy field in the form of electromagnetic waves, so photons come also from an electromagnetic field energy in nature. During the electron jump process, the photons produced by the exchange between $\langle TS_B \rangle_e$ and $\langle TS_B \rangle_{em}$ deduce the wave-particle duality of light.

4.3. QUANTUM ENTANGLEMENT OF THE ELECTRONS IN COVALENT BOND OF H_2

The systematic error that occurs between the photon's spectrum of the H_2 molecule and the calculated results of the Schrodinger equation arises from the change of microscopic states $\Delta S_{B,e} \neq 0$ that can't be calculated by Schrodinger equation when the molecular electron pair on the orbit is formed. It can't be explained by simple "commutants", but can help us to further demonstrate that photon radiation from the electron momentum space configuration potential $\Delta(TS_B)_e$. This can further prove that there is a potential energy $(TS_B)_e$ in the electron movement on the orbits of the atoms and molecules.

H_2^+ hydrogen ion molecular is the simplest molecules, only with two hydrogen nuclei and one outer electron. In 1927, that method of quantum mechanics was used to calculate the energy of electron on molecule's orbit in hydrogen molecules. Although the variation methods and iterative algorithms based on various combinations for all possible electron orbits of hydrogen atom have been developed, and the calculated results can infinitely approximate the actual spectral values, then these algorithms are more like fitting method with experiments. These methods are far from being as clean as using the Schrodinger equation to solve the spectrum of hydrogen atom, which is no doubt that the physical concept used by these methods is wrong.

Relative electron motion, atomic nucleus motion velocity is very slow and can be negligible, so the Schrodinger equation of hydrogen ions molecular approximated by Born-Oppenheimer is a good starting equation [5].

$$\begin{aligned}&\hat{H}\psi\left(r_1,R_A,R_B\right)=E\psi\left(r_1,R_A,R_B\right)\\&\hat{H}=-\frac{\hbar^2}{2M}\left(\nabla_A^2+\nabla_B^2\right)-\frac{\hbar^2}{2m_e}\nabla_e^2-\frac{e^2}{4\pi\varepsilon_0 r_{1A}}-\frac{e^2}{4\pi\varepsilon_0 r_{1B}}+\frac{e^2}{4\pi\varepsilon_0 R}\end{aligned}\qquad(4.11)$$

In the potential energy of Hamilton, the Coulomb attraction between electrons and nucleus and the Coulomb repulsive action on the R distance for two nuclei are included. Namely there is follow equation

$$\begin{aligned}&\hat{H}_e\psi_e=E_e\psi_e\\&\hat{H}_e=-\frac{\hbar^2}{2m_e}\nabla_e^2-\frac{e^2}{4\pi\varepsilon_0 r_{1A}}-\frac{e^2}{4\pi\varepsilon_0 r_{1B}}+\frac{e^2}{4\pi\varepsilon_0 R}\end{aligned}\qquad(4.12)$$

It is very important that the equation logically gives the wave function ψ_e that is constructed by the linear combination of ground state of the electron in atom A and B.

$$\psi_\pm=c_1 1S_A\pm c_2 1S_B\qquad(4.13)$$

It represents the undetermined wave function of the electron orbits (LCAO molecular orbits) in the molecules. The wave function for the + sign satisfies the normalized condition $\int d\tau \Psi_+^* \Psi_+ = 1$. For a given R value of the distance between two nuclei, and considering the symmetry with $c = c_1 = c_2$, for the electrons in H_2^+, the minimum energy value of H_2^+ molecular on ground state of 1.7789-E is obtained at inter-nuclear distance R = 132 pm. The actual spectral value corresponding to the lowest molecule orbit is 2.7928eV at the nuclear spacing R = 106 pm.

Why does the equation with great precision for H atom have an error up to 36.3% in calculating the orbital energy of H_2^+ molecule? This is related to the change of $T\Delta S_{B,\,e}$. According to the Pauli incompatibility principle, two atoms with different spin quantum numbers of electron can form a covalent bond electron orbit, which leads to the change of Boltzmann entropy of electron $\Delta S_{b,\,e}$. There is all no such potential energy term in $\hat{H}$ for H atom and in (4.12) for H_2^+ or H_2 molecule. So, the calculation result for H atom without such a change is very different from those for H_2^+ or H_2 molecule with such a change.

H_2^+ consists of a hydrogen atom and a hydrogen ion. One has one electron and other does a zero electron to make up the molecular electron cloud orbit. Before the formation of the molecule, the number of electron microscopic states for the possible position of one space orbit and possible spatial degeneracy states can be written as

$$
\begin{aligned}
W_{H,e} &= \frac{g_i!}{n_i!(g_i - n_i)!} = \frac{1!}{1!(1-1)!} = 1 \\
W_{H^+,e} &= \frac{g_i!}{n_i!(g_i - n_i)!} = \frac{0!}{0!(0-0)!} = 1
\end{aligned}
\tag{4.14}
$$

The total number of microscopic states is equal to 2. After the molecule is formed, the microscopic state number of an electron on its orbit for its possible spatial positions numbers and degeneracy state numbers is equal to 1.

$$W_{H_2^+,e} = \frac{g_i!}{n_i!(g_i - n_i)!} = 1 \tag{4.15}$$

Therefore, when the molecule is formed, the change of microscopic state number for the electron cloud space as follows

$$\Delta S_{B,e} = \ln \frac{W_{H_2^+}}{W_H + W_{H^+}} = \ln \frac{1}{2} < 0 \tag{4.16}$$

The calculation of $T_e \Delta S_{B,e}$ requires the average rotation kinetic energy $T_{H_2^+,e}$ of electron around the nucleus on the ground state of molecule. It is assumed that Virial theorem can still be applied to the molecule system. That is, there is still the following relationship between the kinetic energy on the molecule's orbit and its potential energy under the action of Coulomb force.

$$T_{H_2^+,e} = \frac{1}{2}\langle p\not{V}\rangle_e \tag{4.17}$$

For The electron cloud, the interaction invariance principle $\langle p\not{V}\rangle_e = \langle TS_B \rangle_e$ is still valid. The electromagnetic spectra of atom and molecules correspond to the change of the potential energy for momentum configuration space of electrons. Therefore, the above orbital electron kinetic energy can be calculated from the spectrum of the ground state for H_2^+ molecule. Namely

$$\Delta \langle TS_B \rangle_e = 2.7928eV = 2T_e \tag{4.18}$$

The correct kinetic energy value of electrons in the molecule can be obtained. It can also be calculated that the change of partial $\langle TS_B \rangle_{H_2^+,e}$ potential energy caused by the change $\Delta S_{B,e}$ of microscopic state number because an electron is shared by a double nucleus. Namely

$$[T_e(\Delta S_{B,e})]_{H+H^+ \to H_2^+} = \frac{1}{2} 2.7928eV \ln\frac{1}{2} = -0.9679eV \qquad (4.19)$$

The $\hat{H}_e$ in the equation (4.11) only calculates the kinetic energy and Coulomb force potential energy of electron in the molecular orbit cloud. And the total energy obtained from this Schrodinger equation is just equal to partial potential energy $\left\langle S_{B,e_{H+H^+}} \Delta T_e \right\rangle_{H+H^+ \to H_2^+}$ to correspond the spectral energy. The molecular Schrodinger equation does not have a term related to $\Delta S_{B,e,H+H^+ \to H_2^+}$ so that the solution of this equation could not contain above energy in (4.19). The total energy calculated by (4.11) should have the following relationship

$$\hat{H}_e = [\left\langle S_{B,e_{H+H^+}} \Delta T_e \right\rangle]_{H+H^+ \to H_2^+} = -1.7789eV \qquad (4.20)$$

It has been discussed earlier that there is potential energy exchange between $\langle TS_B \rangle_e$ and $\langle TS_B \rangle_{em}$ when photons are generated. Therefore, the spectral energy can be recorded as following $\Delta\langle TS_B \rangle_e$

$$\Delta[\langle TS_B \rangle_e]_{H+H^+ \to H_2^+} = -0.9679eV - 1.7789eV = -2.7468eV \quad (4.21)$$

The error from the spectra values was only 1.65%.

The Coulomb force potential energy calculated by the Schrodinger equation and the spectral value corresponding kinetic energy of orbit electron is the most credible data, so that the accurate results can be obtained. It actually proves the following relationship

$$\Delta\langle pV\rangle_{e,H+H^+\to H_2^+} = \Delta[\langle TS_B\rangle_e]_{H+H^+\to H_2^+} = \Delta\left[\langle TS_B\rangle_{em}\right]_{H+H^+\to H_2^+} \quad (4.22)$$

These results also show that 1. Born-Oppenheimer approximation is accurate enough; 2. There is a change $\Delta S_{B,e}$ in the formation of molecular orbit of electrons so they do have $\langle TS_B\rangle_e$ potential energy; 3. Virial theorem is still suitable for hydrogen molecules. 4. The principle of interaction invariance for the formation of molecule orbit still has $\langle pV\rangle_{e,H+H^+\to H_2^+} = \langle TS_B\rangle_{e,H+H^+\to H_2^+}$ 5. Photons are still the commutant for electron on orbit between the Coulomb force and electromagnetic interaction. These results can be verified by repeating the above calculations with the data of H_2 molecule.

The ground state dissociation energy of the hydrogen molecule calculated by the Hartree-London method is: the lowest hydrogen molecule ground state total energy at distance R = 87pm is 303kJmol^{-1} (-3.139eV), correspond to the spectrum value is equal to -4.7449eV, and calculation error is nearly 34.3.3%. The reason is also that the change of $\Delta\langle S_B\rangle_e$ is not taken into the theoretical calculation. Two electrons with different spin quantum number can share one electron cloud orbit when hydrogen molecules are formed. After the formation of hydrogen molecule, the state number of possible particles n_i = 2 corresponding to number of possible spatial position for two electrons in the same electron orbit cloud. According to the Pauli incompatibility principle, the possible number of degeneracy states $g_i = 2$. Therefore, there is the change of Boltzmann entropy for electrons in H_2 molecular orbit from respectively atom cloud orbit for two H atoms as follow

$$W_{H_2,e} = \frac{g_i!}{n_i!(g_i - n_i)!} = \frac{2!}{2!(2-2)!} = 1 \tag{4.23}$$

$$[\Delta S_{B,e}]_{H+H\to H_2} = \ln\frac{W_{H_2}}{W_H + W_H} = \ln\frac{1}{2} \tag{4.24}$$

From Virial theorem,

$$[T_e]_{H_2} = \frac{1}{2}[(p\not{V})_e]_{H+H\to H_2} = \frac{1}{2}\left[\langle TS_B\rangle_e\right]_{H+H\to H_2} \tag{4.25}$$

Considering the photon as commutant, its energy corresponds to $\Delta\langle TS_B\rangle_e$, so that the kinetic energy of electrons on molecular orbit cloud and partial potential energy $\left[T_e\langle\Delta S_B\rangle_e\right]_{H+H\to H_2}$ can be calculated by spectrum energy.

$$[T_e(\Delta S_{B,e})]_{H+H^+\to H_2^+} = \frac{1}{2}4.7449eV\ln\frac{1}{2} = -1.6413eV \tag{4.26}$$

The total energy calculated by Schrodinger equation is the sum of the potential energy of Coulomb force and the kinetic energy of molecular in orbit. It means that two electrons are pulled to the molecular orbit by Coulomb action and also move with the kinetic energy on molecular orbit. So, there is a different in kinetic energy between molecular and atomic orbits. Then the two electrons do not fuse into a microscopic state. So, this part energy corresponds to the $\left[S_e\Delta\langle T_e\rangle_{H+H\to H_2}\right]$ due to without $\Delta\langle S_e\rangle_{H+H\to H_2}$.

$$\hat{H}_{H_2} = [T_e]_{H_2} + \Delta\langle pV\!\!\!/ \rangle_{e,H+H\to H_2} = [S_{B,e_{H+H^+}} \Delta T_e]_{H+H\to H_2} = -3.1391eV \tag{4.27}$$

The momentum space configuration potential energy of electrons on orbit in H_2 molecular for Coulomb action should be the sum of (4.26) and (4.27), namely

$$\Delta[\langle TS_B\rangle_e]_{H+H^+\to H_2^+} = -1.614eV - 3.1391eV = -4.7760eV \tag{4.28}$$

It is equal to the change $\Delta\left[\langle TS_B\rangle_{em}\right]_{H+H\to H_2}$ for the electromagnetic interaction, so it is the energy of the photon. Its error with the spectral value is only 0.665%.

This proves that the above five results are still true forming hydrogen molecule. There is also invariance under Coulomb force

$$\langle pV\!\!\!/ \rangle_{e,H+H^+\to H_2} = \langle TS_B\rangle_{e,H+H^+\to H_2} \tag{4.29}$$

The above analysis shows that the Born-Oppenheimer approximation is reasonable, and the Hartree-London method for finding the linear combination of the wave function by the variation method is also accurate enough. There is no any reason for linear combination with 2s and or 2p wave functions. One electron can't appear on all possible orbits, after all the orbits are separate, with different shapes and average statistic distance. One electron can have only one microscopic state even for the hybrid orbit, while an electron microscopic state can only have two electrons with different spins quantum number. Therefore, the algorithm is only an experimental fitting even if the algorithm is completely consistent with the experimental value, and has still no physical significance.

The first to be discussed for the above results is the Virial theorem. The kinetic energy of electrons in molecules also satisfies the Virial theorem. It was also confirmed that there was also a "commutant" in the formation of the molecule.

The molecular covalent bond formation process dramatically further demonstrates that there really exists the potential energy $\langle TS_B \rangle_e$ of the electron cloud in molecules and atoms. This demonstrates that this potential energy is also present in the motion of the quantum behavior, not only in the classical mechanics.

The H_2 molecule is more stable than the H_2^+ molecule. Not only in that there is a lower ground state potential but also a closer distance between protons. The Boltzmann entropy reduction of electrons at the time of molecular orbital formation is a spontaneous process. This spontaneous tendency is sufficient to overcome the repulsion between hydrogen nuclei. It can also be compatible two electrons with different spin quantum number in the same orbit. The spontaneous tendency $\Delta S_{B,e} < 0$ is the most fundamental driving action for outer electron recombination. The most important conclusions to be drawn from the discussion of the H_2 molecule are: molecular covalent bond formation is a process in which Boltzmann entropy spontaneously decreases.

This discussion also shows that the concept of spatial degeneracy state in local off-domain statistics can also be used in the probability space of quantum mechanics. Therefore, the existence of $\langle TS_B \rangle$ potential energy for particle motion in quantum mechanics is helpful to understand some inexplicable quantum behaviors, such as wave-particle duality, uncertainty principle, commutant and so on.

Chapter 5

ORDERED LIVES ARE PRODUCED SPONTANEOUSLY

Friedrich Engels is written in "Dialectics of Nature": "proteins are the material basis of life, and life is the form of the proteins exist." this is the expression most pertinent to" life so far. Although one hundred years of life science has made tremendous progress, this expression remains unsound. The idea is really very powerful!

In the age of knowledge explosion, today's life science is rapidly advancing, looking back at this basic "life" proposition, it is still blinded: how highly ordered life evolves from "disordered chaos". Implicit here is the definition of life in the book "What is life" by Heisenberg. I.e., the most fundamental feature of life is considered "order". This stems from the recognition that thermodynamic systems handle random chaotic thermal motion. But in the concept of "order" in momentum space as described previously; the universe is ordered from the beginning. Life is only an order of a special form of movement. It can be presumed that: life should be an ordered assemblage of some movement sets about hydrocarbon chain polymer molecules and ionic that can be replicated and regenerated. This involves Coulomb force and electromagnetic interactions in chemical bonds (including hydrogen bonding and ionic charging) and intermolecular interactions (e.g., a particular form of dispersion force-emptying force). The rotational momentum reserved by long chains in polymer molecules brings infinite space configurations, so that the number of the momentum space configurations formed by the polymer molecules is huge. It is this seemingly chaotic but enormous possibility that provides the possibility of spiraling, folding and assembling structures, which in turn form the basis for orderly life movement. Therefore, life should be the existence just as the universe with inherent symmetric and order without any different. The order of life should be the grounds and confidence that we deeply believe that the order in the universe is ubiquitous. And not be the opposite. This in itself can give life sciences a more accurate framework. It is possible to derive some new starting points from propositions of how life spontaneously arises. In life activities, minimization of free energy is equivalent to spontaneous reduction of Boltzmann entropy in the process; this makes it easier for us to understand life activities.

Force invariance, $p\not{V} = TS_B$, which is also the most basic rule for life activities. It can help us to better explain the intrinsic driving force of life orderliness.

Our discussion is limited to whether life is spontaneously produced, not to the life activity itself. Only some links of cell formation will be discussed below, from which the simplest rules to follow in the process of complex life generation are read. This is expected to help understand the vast number of new findings emerging in life sciences.

5.1. C, H Compound with Amino Acids, Long-Chain Polymer Molecules

When people are confused about the origin of life, the people ask for help from the spirit and naturally think of coming and going beyond the sky, and intuitionally make people look at starry sky. We also really observe the birth of some molecular dark cloud in the universe, which is the H_2 molecular cloud changing from H atoms. They appear in a ring-like structure around the dark cloud of dust [6]. It is self-evident that the significance of the evolution from H atoms to H molecules is observed in interplanetary dusts. More than 140 molecules were also found in space in succession. Formaldehyde was found in the interplanetary gas in 1969, followed by ethanol, acetic acid and glycolal (the simplest sugar molecule) and glycine in 2003. Liquid methane and ethane are also found in large quantities in Saturn, suggesting that nature can spontaneously produce ethane molecules, regardless of the reaction process. The oil and gas on earth may have been produced with the earth and does not need to accumulate the organism in the start over a long geological age. The discovery of cyano-ethylene in Titan provides a very important clue to the problem of spontaneously producing life [7].

90% of mass in the universe is occupied by H atoms. Helium 4, which contains 2 protons and 2 neutrons, is also extremely stable, accounting for 9% of the universe mass, and all other elements only account for 1%. Whereas the C12 atom, which has an integral multiple of the helium 4 nucleus, has a stable core structure, it is a natural choice for a molecule consisting of C and H to have a stable core structure. The elements before Ca40, such as Oxygen O16, Neon Ne20, Magnesium Mg 24, Silicon Si 28, and Sulfur S 32, all have stable core structures. The C, O, N, P elements have very active reactivity that is depended on the electron numbers at most outer orbit, and the atoms of these elements are involved in the construction of living substances (including proteins, DNA, RNA, phospholipids, lipopolysaccharides, peptidoglycans and glycogen) all are natural selection processes. Hydrocarbon, such as Polycyclic Aromatic Hydrocarbons and their derivatives, can find corresponding mid-infrared spectra in space. It is presumed that these hydrocarbons can be formed in the cosmic space interplanetary medium under certain reaction conditions [8]. These evidences indicate that there is no difference between life originating in the universe or at earth, which is originally one corner of the universe and earth only is just a sky of outside the sky. It is important that all of these life-forming molecules are polymers consisting of covalent bonds without exception. This is more closely related to the spontaneous decrease in Boltzmann entropy to form covalent bonds.

5.1.1. C-H Bond in Methane

The discussion of the living matter must involve the formation of the basic C-H and C-C covalent bonds. Analysis of Boltzmann entropy change during the formation of covalent bonds in the C-H bonds of methane molecules and the C-C bonds of ethane molecules can help illustrate the spontaneity of production of long chain C-H polymer molecules.

The carbon atom has 4 outer electrons respectively located on 2S and 2P orbit, and the number of the outermost electron micro-state is 3 electrons on the 2P orbit. The 4 electrons of most outside orbit would form four SP_3 hybridization orbits before forming a molecular orbit with the H atom, which are four entangled microscopic quantum states. The four hydrogen atoms have 4 microscopic states for four electrons. There are 7 initial microscopic states in total with the carbon atom having 3 outer 2P states and four 1S states for four hydrogen atoms. Four electron pairs in covalent bond of the generated methane molecule should be

$$\Delta\langle S_B\rangle_{e,4H+C\to CH_4} = \ln\frac{4+4}{4+3}\frac{4}{4+4} = \ln\frac{4}{7} < 0 \tag{5.1}$$

If two H_2 molecules start to react, 4 electrons only have 2 covalent bond microscopic states, and the microscopic state number change in the reaction process is

$$\Delta\langle S_B\rangle_{e,2H_2+C\to 4H+C\to CH_4} = \ln\frac{4+4}{2+3}\frac{4}{4+4} = \ln\frac{4}{5} < 0 \tag{5.2}$$

Therefore, the reaction of generating methane is a spontaneous process no matter opening from the H atom or the H_2 molecule and the reaction with a carbon atom reach the reaction condition and exceed the reaction potential.

5.1.2. C-C Bond in Ethane

The C-C bond in ethane is the basic unit of a long chain C molecule that all living matter possesses. The free rotation of the diatomic molecular C-C bond is a very important feature. Long -C-C- chains produce countless spatial conformations, even though not necessarily with arbitrary rotations on the atomic scale, and on the scale of the Kuhn segment, new microscopic states can be produced by rotation. The momentum configuration of electron cloud definitized by the C-H bond rotating along with the C-C bond will form a new rotating micro-state related to original -C-C- bond. This uncertainty is essential to the adaptability of the life for living condition in nature. But the living body is deterministic. Thus, discussing C-C chain generation can help understand how life balances certainty and adaptability based on the links at the most basic structure unit.

The rigid bond of the diatomic molecule has two quantum numbers, j and m. The degeneracy number of each energy level is equal to $2j+1$ for a quantum state. There is no zero angular momentum. Therefore, the rigid bond of a diatomic molecule will even have a lot of possible microscopic state due to rotation. This is in contrast to the change in the number of microscopic states caused by the spin quantum number of the outer electrons during bonding. The number of microscopic states of the long-chain -C-C- bond structured living matter can be very large. Therefore, how does the order and certainty of life complete in a spontaneous process, and C-C bond of ethane is a good problem.

Assuming that two carbon atoms and six hydrogen atoms can generate one ethane molecules C_2H_6 no matter what reaction is carried out and what reaction conditions are needed, tree electrons in 2P orbit at the outermost layer of each carbon atom are hybridized with one electron on 2S orbit to form 4 SP_3 hybridized orbits, then among them, three electron form C-H covalent bonds with three electrons in three hydrogen atoms, and the remainder electrons on two carbon atoms form one C-C covalent bond. And 7 σ covalent bonds are formed. Through the change of the number of outer electron microscopic states to produce the ethane molecules from the carbon and hydrogen atoms the Boltzmann entropy change is thus

$$\Delta S_{B,6H+2C\rightarrow C_2H_6} = \ln\frac{6+2\times4}{6+2\times3}\frac{7}{6+2\times4} = \ln\frac{7}{12} \qquad (5.3)$$

But the hydrogenation of acetylene or ethylene to ethane is carried out under the conditions of high temperature and high pressure when the catalysis of Ni or Pd is carried out. In this process the energy of π bond in the double bond of ethylene are lower than the energy of σ bond and π bond is easily broken, but why is the reaction quite severe?

There are 6 microscopic states of outer layer electrons in an ethylene, which are composed of 5 σ bonds and one π bond. Two electrons of a hydrogen molecule have only one microscopic state in a covalent bond. So, there are seven microscopic states before the reaction. But after turning into ethane there are still seven microscopic states. The Boltzmann entropy change is as following during reaction process

$$\Delta S_{B,C_2H_2+H_2\rightarrow C_3H_6} = \ln\frac{7}{6+1} = 0 \qquad (5.6)$$

And this is not a spontaneous reaction process.

From the fixed C = C double bond in the ethylene which can't be related forms C-C single bond in ethane. A number of spatial rotational conformers are suddenly developed. This reaction has to be carried out under extremely difficult conditions with the aid of catalysts. The catalyst has the function of changing hydrogen molecules into hydrogen atoms, and the corresponding reaction of two hydrogen atoms has

$$\Delta S_{B,C_2H_2+2H\rightarrow C_3H_6} = \ln\frac{7}{6+2} = \ln\frac{7}{8} < 0 \qquad (5.9)$$

The reaction could proceed spontaneously. But still suffer from increased microscopic states due to spatial rotational conformation.

The ethane found in Titan indicates that the reaction process for producing ethane may be by many ways, and the molecule can be generated spontaneously in nature.

5.1.3. Long-Chain Polymer Molecule

The double bond C = C of ethylene is opened and polymerized as long chain molecule -[-C-C-]$_n$-of polyethylene with single bond through addition reaction. After the π bond in the double bond is opened, each of two electron are respectively formed into two σ bond, so that half microscopic state for each electron is given and have following Boltzmann entropy change

$$\Delta S_{B,nCH_2=CH_2 \to \{CH_2-CH_2\}_n} = \ln \frac{n \times 2 \times \frac{1}{2}}{n \times 1} = 0 \tag{5.10}$$

The reaction is also not a spontaneous process. Boltzmann entropy change can be used to explain why this addition reaction requires such severe reaction conditions as catalysts and high temperatures and pressures. In life activity, even under the catalysis of enzyme, there are not reaction condition for forming -C-C- bond; this long chain molecules are produced via other reaction pathways.

All protein molecules are linked by peptide bonds consisting of amino -NH_2 and carboxyl -COOH among twenty amino acids. The -OH group in deoxyribonucleotide and phosphate acid through esterification reaction forms the phosphodiester bond, so that several different deoxyribose nucleotides are linked by the phosphodiester bond to long-chain molecule of DNA. This indicates that all living matter is composed of carbon and hydrogen atoms, but only the frame. The C-C bond and the C-H bond are molecular platforms for building life substances, or molecular reaction platforms. The number of microscopic states generated by the rotation of C-C bond is related to the number of carbon atoms. Deterministic living organisms require some structures in the molecule that can prevent the creation of rotational conformation and produce lot of micro-states. The structures of, for example, the group of hexagonal heterocycles and bicycle-group in the primary structure of the polymer, or the occurrence of specific sequences, for example in the HP folding model, or the side groups (residues) of specific spatial helical arrangements, hydrogen or secondary bonds of intermolecular or intra-molecular, all serve to prevent formation the microscopic state produced by the rotation of this C-C single bond.

5.1.4. –NH_2 and –COOH Group in Peptide Bond

In that condensation polymerization reaction of –NH and –COOH groups, a –COOH and a water molecule are generated by the respective electron pair of NH and OH. Their respective spin electron pair microstates become electron pair microstates in two new molecules, generally there is $\Delta S_B = 0$. However, since the reaction starting from two amino acids molecules, for one dipeptide molecule and one water molecule, the variation of the microscopic state number changes of electron spins is respectively as $\Delta S_B = \ln\frac{1}{2} < 0$. In fact, the poly-condensation of amino acids is a spontaneous process.

The acrylonitrile molecule is very active compound and can react with water to form amino and carboxyl. And the processes are proved to be spontaneous processes with Boltzmann entropy reduction. It can also form a heterocyclic ring with a -CN group by the double bond reaction. And thus is an important substrate for the production of living matter. After the compound is detected in Titan, the NASA will implement a space plan for detecting whether a live matter is there or not on Titan in the near future. The basic framework of proteins is framed by peptide bonds. Amino and carboxyl groups are key groups for constructing proteins, and proteins linked by peptide bonds can spontaneously create.

5.1.5. Phosphodiester Bond in DNA

The phosphorus is the sixth most abundant element on earth. Phosphorus is an important element in living matter, and is a derivative of phosphate, namely, calcium phosphate in bone and teeth, lecithin constituting cell membranes, phosphodiester bond linking DNA and RNA, and high-energy phosphate bond in ATP energy unit. Phosphorus was first observed in the deep space in 2009, namely in the residual of the Big Bang, Cassiopeia A.

And the ion-state oxygen can be generated in the nuclear fusion of the large-mass fixed star, and the molecular oxygen can be observed after cooling. Phosphorus and oxygen are two major elements found in deep space. The formation of phosphodiester bonds is only discussed here.

Phosphorus P is more active than nitrogen N and is very easy to oxidize to generate oxide which is hydrolyzed into phosphoric acid. This phosphorus-oxygen tetrahedron has tree SP_3 hybrid orbits forming three σ bonds with the outer electrons of oxygen, then forming three hydroxyl groups with H^+. The other is a triple bond, and there is no change in the reaction. It is composed of σ bond of SP_3, (phosphorus atom lone pair electron matches oxygen atom empty orbit); and π double bond overlap. A π bond is a double bond consisting of the d_{XZ} and d_{YZ} empty orbit on the phosphorus atom and two isolated electrons on the P orbits of two oxygen atoms.

The hydroxyl groups on the 3', and 5', in five carbon rings of ribose or deoxyribose are esterified with the H^+ groups on the phosphate to form phosphodiester bonds, which link into long-chain DNA or RNA molecules and 2 water molecules. In this process, on a five carbons ring sugar molecule, the outer electron microstructure is broken from the σ bond, with two opened -OH bonds on the phosphate, involving 4 spin microstates of electron pair. There are to be formed two new electron microscopic states in two water molecules and two new -C-O electron spin microscopic states on one phosphodiester bond. The overall electron spin microscopic state is also changed $\Delta S_B = 0$ as is peptide bond formation. However, for the formation of DNA molecules and water molecules, respectively, there are

$$\Delta S_B = \ln\frac{1}{2} < 0 \tag{5.11}$$

The lipidation process to form the phosphodiester bonds of DNA and RNA is a spontaneous process. The space arrangement of deoxyribonucleotides residues on the phosphodiester bonds is compact, the microstates caused by the rotating quantum number between long chains is not generated, five hydrogen bonds formed between A-T and G-C are also like-covalent bonds, and the generation of the hydrogen bonds has also the spontaneity of Boltzmann entropy reduction. And the number of microcosmic states of spatial rotation conformational isomers is more reduced; the possible number of microscopic states of deoxyribonucleotides on DNA and ribonucleotides on RNA in the sequence of linkage, number of permutations in the genetic code between strands is also reduced.

The spontaneous generation of ordered life from atomic or molecular morphology of elements by the spontaneity of hydrocarbon, protein and DNA production reactions should no longer be questionable. Of course, there are a series of evolutionary processes that become "live" life forms, and the "live" nature of these life forms will be analyzed below.

5.2. Hydrophobicity, Membrane and Equilibrium State

The hydrocarbon C-C bond is not the main link mode of tectonic living matter, but the hydrophobicity of the various hydrocarbons is very special. Water is the essential condition for life. The hydrophobicity is the premise and basis for the individual body to distinguish from water environment and one another from each other. Therefore, it can also be said that the hydrophobicity of hydrocarbon is one of the most primitive drivers generating life. It is the driving force for generating membrane. And the imbalance between membranes is the basic driving potential of living activity.

5.2.1. Spontaneous Formation of Droplets for Hydrophobic Media

Trace amounts of hydrocarbons do not be dissolved in water, form droplets in water and gathers to be larger enough. If viewed from the momentum space configuration of the liquid phase, the hydrophobic droplets in the surrounding liquid water can be considered to be in a space free of wall without constraint, i.e., can't regard to the shear stress imparted by the container walls to the droplets (such action is discussed in the phase change problem). Such hydrophobic droplets can therefore also be discussed in terms of a like wall-free space ideal gas model. The intermolecular transfer of the rotational momentum of molecules is the intermolecular interaction force. The action force on continuous medium level is referred to as shear stress. Now the momentum transferred in liquid is rotational momentum of mass point rather than translational momentum (there may be some vibration momentum corresponding to perturbations). Since the molecular weight of the long-chain CH compound is much larger than that of water molecules and the molecular chain is much longer, the rotational inertia of the long-chain molecule is larger and the rotational momentum is much smaller than that of the water molecules at the same temperature. Thus, a spherical distribution with a small rotational momentum centered on the CH compound in the water is the liquid drop. There may be aggregation (or complexes) of water molecules connected by hydrogen bond with greater rotational momentum around them, and smaller water complexes and to other positive or negative ion hydrates further on the outside. They are distributed in the space in a spherical symmetrical mode along the radial direction according to the magnitude of the rotating momentum. The mixture is broken into emulsion by stirring, is not stable by a surfactant, and can still spontaneously gather into a larger drop. Until the spherical symmetric surface of the liquid drop is split by additional shear forces due to deformation. The tendency of the hydrophobic macromolecule to aggregate into a liquid drop does not have to be explained by surface tension. Furthermore, many hydrophobic compounds are spatially arranged in the droplet according to the magnitude of the rotational momentum generated by the magnitude of the respective moments of inertia. This makes it possible to assemble a particular membrane-forming molecule.

5.2.2. Depletion Force

The effect of the spatial distribution of momentum according to the size for the polymer with very regular spatial configuration is the depletion force discussed in the biophysics. High momentum particles of small molecules are pushed out of the middle of the polymer. The tight alignment also causes the dispersion forces (Van der Waals forces or internal pressure) between the polymers to become very large. These polymers are assembled into the poly-gel, such as cell membrane.

5.2.3. Phospholipid and Spontaneously Generation of Cell Membranes

When hydrophilic groups are present on CH compounds by chemical reactions, hydrocarbons become amphiphile, also known as surfactants, for phospholipid it generates a kind of cell membranes molecule. It can dissolve the CH compound. Furthermore, the hydrophilic group can form a hydrogen bond with a water molecule, and thus it is very important that the amphiphilic molecule occurs during the formation of life. Hydrogen bonds formed between hydrophilic groups and water molecules (experiments have confirmed that hydrogen bonds are a kind of weak covalent bonds), although Coulomb actions are weak relative to covalent bonds. But the same is true with respect to the formation of the common orbit of electrons pair, as is the spontaneous process by the reduction of Boltzmann entropy.

The phospholipid molecule constructed from element C, H, O, N and P is the amphipathic molecule. Choline with hydrophilic amino group takes part in aliphatic reaction with two alkyl hydrophobic tails through phosphate acid to form lecithin. (Spontaneity of the lipophilicity of the reaction process has been discussed) Similar reactions can produce a family of phospholipid molecules. They are molecular materials that assemble the membranes of various cells and organelles.

The aggregation and assembly of amphiphiles in aqueous solution into a membrane depends on the size of the hydrophilic group and the geometry and ratio of the hydrophobic groups' length to the head size of the hydrophilic group. The size of the head of the hydrophilic group is large enough, the hydrophobic group is slender enough to obtain a spherical radius R, and the tail of the hydrophobic group is completely wrapped in the sphere of which the hydrophilic group is spherical. The number of hydrogen bonds is the largest on the sphere's surface. The maximum Boltzmann entropy reduction is obtained in the minimum space.

Such spherical surfaces of micelle have a tendency to contract the surface compactly. Smaller molecules or ions can be pushed out of the gaps between phosphatides. So that Van der Waals force among phosphatide molecules is increased, and the liquid-phase intermolecular interaction is a tangential dispersion force generated by the rotation momentum when the molecules contact or collide, and the dispersion force can rub the spheres more densely. Like the effect of rubbing rice flour by hands of Chinese people. The film will thus become very dense. A membrane is a collection of molecules assembled with less dispersive forces than hydrogen bond, and a kind of poly-gel (or poly-polymer).

However, when the geometric dimensions of the head and tail of the molecule are not well matched, for example, an excessively large head and an excessively short tail can cause space which cannot be occupied by phospholipid molecules but by water molecules in a spherical space coated by the membrane. And new condition for contact between hydrophobic molecule and water can cause entropy increase. The membrane structure of double layer phospholipid may also form hydrogen bonds with water molecules inside the membrane. It can accommodate more general conditioned coating tasks. When a bilayer membrane molecule is formed, more phosphate esterification reaction is also spontaneously generated in nature.

This vesicle, formed by the phospholipid bilayer membrane, is a decisive step in the formation of life. It provides a stable interface for hydrophobic CH compounds; it is a prerequisite for the generation of life activities and life evolutions; the life individual who makes evolution change has a boundary layer between yes and no.

5.2.4. Membrane and Chemical Potential Equilibrium

Because of the membrane, it seems to be in a life far from the thermodynamic equilibrium state, but maintains its own thermodynamic equilibrium state in the space isolated by a membrane. Namely the membrane separates chemical potential difference (temperature difference, pressure difference and concentration difference) inside and outside cells. Respective systems on two sides of the membrane can be still seen as thermodynamic equilibrium systems, so that thermodynamic imbalance between two sides of the membrane is the original driving action for constructing life activity. So, we can not only discuss the life phenomenon in equilibrium state thermodynamics, but also keeps the driving potential of life.

The difference of life activities of each part among life bodies can generate temperature difference, but the difference can be achieved constant value or keep it within a reasonable fluctuation range by means of respective molecular motion inside and outside the membrane or the flow of air and some body fluid. When the objects with different temperature are contacted, heat flows from a high-temperature object to a low-temperature object, and Boltzmann entropy is spontaneously reduced, so that each part has the capability of tending to temperature balance.

The pressure difference between the inside and outside of the membrane can be balanced by the tension of the membrane. The thermodynamic equilibrium must be satisfied for both sides of the membrane to be maintained. This pressure difference is represented by the osmotic pressure. The osmotic pressure due to the concentration difference is also adjusted by the concentration difference. The concentration difference on both sides of the membrane can be very large; but each side is uniform. Thus, each side maintains the thermodynamic chemical potential in equilibrium state. They are just the most pivotal driving forces for life activities. The change of the membrane tension caused by osmotic pressure is a control signal of a protein channel switch on the membrane and is also a signal for regulating and controlling the concentration.

There may also be a difference in chemical equilibrium constants on both sides of the membrane. This is also the form in which life activities appear. Not only can the ionic balance, reactant and product balance on each side still be maintained. The membrane also maintains the Dannan equilibrium between ions on both sides and C-H compounds, which is the basis of all ion channel regulation mechanisms.

Almost all reaction processes are catalyzed by enzymes. Not only reduces the reaction potential, but also avoids multi-path and reversibility of the reaction. The reaction process is maintained under the living condition of the living body; a series of reactions induced by receptor-ligand matched signal pathways or operons have the certainty and directionality of the reaction, and ensure that the chemical processes in the life activity can be completed under the two states control. When, for example, an extra heart pulse rhythm occurs, this may not be the pathological process required by the life process. Since a large number of proteins, ions are inside cell, the process is not triggered by specific enzymes and will be an unusual reaction pathway. The reaction clock, such as the B-Z reaction, is a nonlinear kinetic behavior. Whereas the periodicity caused by the non-linearity even generates chaos. It is not an essential kinetic process to life activities.

In a living body, there may be a flow of an intra-systemic medium, such as a flow of blood in blood vessels and a flow in lungs and trachea. These are in laminar flow in fluid mechanics, are not in the nonlinear dynamic range, and are far from being considered flows away from equilibrium.

All of these membrane-to-membrane imbalances and flow and chemical reactions occurring in living organisms are not the reason for viewing life activities as far away from thermodynamic equilibrium states.

All processes can be performed in a linear, bi-reversible manner due to the presence of the film. All states can be described by thermodynamic variables. The state is deterministic despite the randomness of the system. Since all regulation is two-state control. The controllability of the process also allows the conditions on both sides of the membrane to be determined.

The discussed heat and mass diffusion processes are also spontaneous processes of Boltzmann entropy reduction. The irreversibility of these processes stems from: microscopic randomness in the determinacy of macro variation, for life activities we have observed that in minimizing this randomness at the microscopic level, living organisms have even evolved to "pack" with vesicles for transport along microtubules, this ordered directional transport greatly reduces the randomness of molecular level in the flow and the irreversibility of the three-dimensional spatial diffusion process by virtue of this randomness. Understanding life activities based on the discussion of thermodynamic equilibrium states is therefore more advantageous than the recognition based on far away from thermodynamic equilibrium states. There are some irreversible processes in the process of life continuation, but it can't be said that irreversibility results in such orderly dissipative structures like life. The molecular motor is driven by energy cost, such as ATP hydrolysis, to realize the reverse gradient transport process of Ca^{+2} and K^{+} ions through membrane, and there is negative entropy flow. But it and the positive concentration difference transport process through the ion protein channel on membrane can be regarded as forming a double 'mass source' heat engine. Thus, as with a normal heat engine, a single negative entropy flow path is not sufficient to consider the entire heat engine system as a "dissipative structure" far away from equilibrium. Instead, the living body is more like a high-efficiency reversible heat engine consisted by a large number of cells that each cell can be regarded as a micro-Carnot cycle heat engine.

This potential energy TS_B with the ability to build new thermodynamic equilibrium states is not ergodic. As previously described, the ergodic process violates $\Delta S_B < 0$.

And according to the invariance of the acting force, the acting force changes the distance space and simultaneously completes corresponding momentum space rearrangement. This spatiotemporal correspondence depends on the force that changes momentum. The life activity is in the continuous medium acting force range. The fastest is the process corresponding to electromagnetic forces in molecule, followed by intermolecular forces, and the slowest is the time scale between continuous media. The time scale of such a process is related to the energy scale. The uncertainty relationship should be satisfied on a quantum scale. The longest time scale is the life span of the individual. Also appears to be positively correlated with the length of the mitochondria. The adjustment process in the life activity is a rather fast targeting process and not a search process.

5.2.5. Channel Inlaying in Membrane

The membrane maintains an imbalance in thermodynamic parameters between the two sides, and the living body is working on this imbalance. Exchange between the two sides of membranes is an important feature of life activities. This exchange is accomplished by means of channels and molecular motors (e.g., ion pumps) formed by protein molecules embedded in the membrane. The membranes are assembled by dispersive forces and the "appendages" on these membranes can be embedded. Therefore, no reaction procedure is required to form membrane protein channels on the membrane, and no process editing difficulties exist. This is just as with patterned carpets, where the difference between the patterning through the copolymerization reaction and the weaving process is too great. Formation of ion channel protein "patterns" on a membrane is not possible by chemical reaction. It has also been demonstrated, in turn, that the membranes are assembled by dispersive forces. Life chooses this process of spontaneous formation that a membrane with ion channels is constructed by the method of assembling with the least number of microscopic states. The number of molecular micro-states that the reaction process will use is much greater than the number of micro-states used to compile how many assembly patterns are used.

The appearance of membranes on the organelles of eukaryotic cells implies the perfection of various functions that various life needs have. When the cells are collected into an organ, the organ membrane is also present outside the organ; the organ is in turn covered by a more complete membrane of the skin into a more advanced living body. In the evolution of life, membranes are also constantly evolving. A multi-level, more rational and directed balancing and adjustment process also occurs.

5.3. Osmotic Pressure and Regulation of Membrane Channels

The various pathways on the cell membrane are very numerous, and the individual modes of regulation and signals of regulation all differ, but the system-regulating signals should be maintained for the whole cell. It is believed that this signal most likely should still be the one that maintains thermodynamic equilibrium in the system: temperature and pressure signal. There will be temperature and pressure differences on both sides of the membrane and between the organs. The original regulatory signal should be the osmotic pressure across the membrane. But the explanation of the method is somewhat strenuous and difficult to understand by used of the principle of minimization of free energy; the osmotic pressure concept has a fundamental role in biophysics; carding-down is necessary.

5.3.1. Crowding and Osmotic Pressure

Many protein molecules and various ions and water molecules are contained in cells. The cells are very crowded. When these different molecules are all mixed enclosed in a membrane, mixing is a spontaneous process $\Delta S_{B,Mix} < 0$. In terms of this change in the membrane, there is an isothermal, isochoric mixing process, according to the force invariance

$$(T\Delta S_B)_{Mix} = (V\Delta p)_{Mix} \tag{5.12}$$

The interaction in solution of Incompressible continuous medium should be the pressure p in continuous medium motion level, according formula (5.12) means that the pressure in the solution and mixtures on the inside of the membrane will become smaller, i.e. $\Delta p_{Mix} < 0$. There is less than pure liquid or solution with lower concentrations on other side of the membrane. Or it can be said that "osmotic negative pressure" appears, which is contrary to textbook-like "common sense" of osmotic pressure elevation. The common sense is indeed a misreading that looks only at the results.

Considering the cells as a system with the surrounding solvent, it is the high pressure on the pure solvent side that allows water molecules to flow into the cells through the gaps in the assembled membrane or the channels on the assembly membrane. The number N of water molecules penetrated into the cells by the surrounding solvent in the process is as follows; namely there is N microscopic states, the N water molecules within cell change after mixing the offset into N' micro state. From $N > N'$ the total infiltration process, the intra-and extra-membrane system Boltzmann entropy change corresponding to the osmotic pressure is generated

$$\left(\Delta S_B\right)_{Osm} = k_B \ln\frac{N'}{N} < 0 \quad \left(N' < N\right) \tag{5.13}$$

So, this osmotic negative pressure occurs spontaneously, and will continue to infiltrate until the solution is infinitely diluted.

If the system has an open boundary, the liquid pressure remains constant on both sides of the membrane, and the system continues to increase in volume due to the penetration of water molecules, appearing at equal pressure $(pd\mathit{V\!\!\!/})_{Osm} > 0$. As shown in the figures on many textbooks, when the liquid level rises due to the increase of the volume, the level of the solution side rises higher than the open level of solvent, for example water. The hydrostatic pressure on the solution side at the permeable membrane rises to equal the pressure on the solvent side, and the osmotic pressure on the solution side ceases to continue the solvent permeation. The above concept of "osmotic negative pressure" is well documented by the volume change of the solution side. The cell is a closed space, when the cell goes through an osmotic negative pressure stage, the membrane expands to generate intra-membrane tension along with the increase of the volume, the intracellular pressure goes the process $(\mathit{V\!\!\!/}dp)_{Osm} > 0$, and meanwhile, the cell membrane tension increases until the membrane tension balances the osmotic pressure of the solution. Take the cell as a system, water molecules continuously enter the cell that is a process of $\Delta(p\mathit{V\!\!\!/})_{in} > 0$, water molecules increase the N' number of microscopic states, and that is a $\Delta(TS_B)_{in} > 0$ process under the isothermal process. It meets force invariability principle for the intracellular action, and the osmotic pressure is represented as elastic tension of the membrane.

In fact, extracellular is also a solution in which water molecules infiltrate into the cell to reduce N'' microscopic states that is a process of Boltzmann entropy reduction for solution in the outside of cell. But there is $N'' > N'$ and the water molecules can still spontaneously infiltrate into the cell until the extracellular solution is dehydrated and the increase in the concentration of the solution will stop from continuing to penetrate into the cell and even reverse osmosis. The limited concentration difference between two sides can be balanced again. A cell put into the clean water will break due to the continuous infiltration of water resulting in infinitely dilute intracellular solution.

Later on, the marker was said to be precisely, the infiltration was due to the spontaneous tendency of systemic dilution, the infiltration produced cell membrane tension. The membrane tension produced $\left(pd\mathcal{V}\right)_{Osm} > 0$. It is sufficient to explain that a large amount of water is delivered to the tree along the lignin cells. The plant cell membrane tension due to the osmotic pressure is much greater than that of the capillary force to indicate the water can rise from the root to a few tens of meters high. Knee cartilage is kind of charged poly-gels, that is a kind of glycosaminoglycan, which also has this infinite dilution capability, sufficient to generate osmotic pressures up to several hundred atmospheres pressure. So that the cartilage cell membrane has great tension and can buffer huge impact force.

5.3.2. The Ion Channel Proteins of the Membrane

Some protein molecules with special space configuration are spontaneously generated in nature, the hollow space scale is equivalent to the special ion scale, and ions can pass through the hollow protein molecules. The presence of such ion channel proteins in water bodies is possible and allowed to occur spontaneously, they are simply a protein molecule with a particular spatial conformation. It is also endocytosed and fused by the phospholipid membrane because the molecule has a hydrophobic moiety (which is aligned in accordance with the hydrophobic moiety in the HP model discussed below for molecular folding), and thus the conjugation of this hydrophobic moiety to the hydrophobic layer in the phospholipid bilayer membrane occurs spontaneously. Also, this combination retains the possibility that more water molecules form hydrogen bonds resulting in more Boltzmann entropy reduction. The process of inlaying ion channels on the membrane is also spontaneous.

5.3.3. The Force-Sensitive Control for Ion Channel on the Membrane

The osmotic pressure directly affects the regulation and control of ion channels on membrane. It involves in the transport of water molecules and ions on the cell membrane, enabling cells to "live".

The numbers of water molecule channel protein embedded on membrane have a statistical property, the number of possibly opened channels is very sensitive to the load force signa, that is membrane tension caused by osmotic pressure. The membrane tension generated by osmotic pressure can administer two-state control, while the osmotic pressure is determined by the ion concentration. The numbers opened ion channel is kept within the range of ion concentration different between inside and outside cell. The two-state control on the mechanics sensitive channel is regulated and controlled by 'osmotic negative pressure' and membrane pressure generated when reaching positive osmotic pressure. Osmotic pressure is the most reliable mechanical signal in living beings. Under the action of membrane tension stress τ generated by osmotic pressure, the spatial momentum configuration on the membrane changes to entropy change $\Delta S_{B,Memb}$. A channel annular stress p_{Memb} is provided on the annular channel $\Delta\left(pd\not{V}\right)_{Memb}$; p_{Memb} is the equivalent line tension f in the membrane channel analytical solution given in document [9].

The stress action on membrane protein channels of water molecule produces two-state control problem is: the control problem is accomplished by the tension τ in the membrane caused by osmotic pressure resulting from mixing and the tension f of the toroidal channel on the membrane to switch channel. Some supplementary explanation is necessary for the above analytical solution.

The f inward negative value is converted into the outward positive value, and the τ value is defined as τ_{crit} for $f = 0$. While $\tau > \tau_{crit}$, namely $f > 0$, f is defined as f_{open}. Vice versa $f < 0$, f is defined as f_{clos} being off. Both all are functions of τ . However, the two functions are different and the variation range is also different. The f_{open} change is fast and f_{clos} slow, depending on the thickness of the film. The normal state is closed state, and the f_{clos} change is limited. When the membrane tension τ becomes large, namely the concentration changes to the point that the membrane thinned needs to be restored again, the channel can be responded quickly, and the water molecule channel is opened. The osmotic pressure of cells is automatically maintained within a reasonable range while the water molecules flow out of cell according to the channel protein own τ_{crit} . Osmotic pressure and τ_{crit} of water protein channel together constitute a two-state regulation and control mechanism that is a very stable negative feedback regulation.

The two-state control is controlled by the spontaneous regulation of osmotic pressure. It not only regulates the flow of water molecules into the cells, but also regulates the outflow of water molecules by opening the water molecule protein channel through the effect on the membrane tension τ .

The transport flux of water molecules through the cell membrane is greater four orders of magnitude than the flux of glucose molecules. And the flux of glucose molecules is greater four orders of magnitude than the flux of ion. The difference between these fluxes is, of course, related to the numbers of corresponding channels. However, the number of water molecular channels is greater than that of various ion channels, which still does not seem to explain the large flux of water molecules on the semi-permeable membrane.

The cell membrane is assembled by intermolecular dispersion force of phospholipid; the structure is much looser than that of the artificial reverse osmosis membrane connected by covalent bonds; semi-permeable membranes used in osmometry are all “dead” biofilms where will not be regulation action in living cells. Obviously, the action of a channel regulation switch on the membrane generated by concentration change in the living body is not generated, and the osmotic effect can still be realized. Therefore, the permeation is mainly not dependent on the water molecule channel to transfer water molecules. The artificial reverse osmosis pressure is operated, and water molecules can pass through a reverse osmosis membrane made of artificial high polymer materials under the condition of positive pressure. Therefore, water molecules can directly penetrate cell membranes under the condition of 'osmotic negative pressure'. Just like the inhalation of a human lung, the whole lung must be inflated under negative pressure. The "osmotic negative pressure" enables the entire membrane to work, which more reasonably explains that the cell membrane has a flux of water molecules that is 8 orders of magnitude larger than the ion channels.

Why are water membrane channel proteins present at the same time? A reasonable explanation would be: the main functions of the water molecule membrane protein pathway are pressure relief and generation of water molecule flow out along the direction of reverse osmotic pressure. In fact, the function of water protein channels only is to expel water molecules. So, as it can to achieve the rebalancing of the ion concentration on both sides of the membrane.

When the membrane tension reaches the upper limit and the ion regulation lags behind, it is necessary to balance the pressure inside and outside the cell and help balance the ion concentration by opening the molecular channel of water. This is also a kind of protective function to the cell membrane. The pressure satisfies the Poisson equation, so the pressure response occurs instantly. Only this transient equilibrium process requires the probability related to the opening number of channel. The water molecules transported directly through the membrane by the osmotic negative pressure are of course not related to the number of water molecular channels, but only to the Donnan equilibrium related to the concentration of ions inside and outside the cell.

5.3.4. Donnan Equilibrium and Ion Channel

If all CH compounds in the membrane are hydrophobic, this destabilizes the phospholipid bilayer membrane structure. The intracellular molecules must be water molecules and hydrophilic organic substances, so that the intracellular molecules have the least total number of microscopic states through hydrogen bond with the double-layer phospholipid. Hydrophilic ionic organic compounds, such as amino acids and polypeptide polymers, can be formed spontaneously.

The high-concentration Na^+ ion environment in seawater, if the Na^+ ion concentration inside the membrane is far less than that Na^+ ion of the environment, water molecules will seep out of the membrane. In order to maintain the water molecules inside the membrane, the high concentration of K^+ than one outside the membrane is needed for intracellular ionized hydrocarbon polymers. So that this not only maintains the balance of dielectric balance in the membrane; but also does the balance of total ion concentration and ion pair Na^+/K^+ between the inner and outer sides of the membrane. It also ensures the bidirectional regulation of the osmotic pressure. The charge balance among the three constitutes the so-called Donnan equilibrium [10].

$$\begin{aligned} (\Delta p)_{osm} &= (z+1)cRT \qquad (c' << c) \\ (\Delta p)_{osm} &= cRT \qquad (c' >> c) \end{aligned} \tag{5.14}$$

Here c' and c are the concentrations of extracellular element ions and intracellular hydrocarbon polymer ions, respectively. The z is the number of elemental ions corresponding to one hydrocarbon polymer molecule. It is the basis for charge rebalancing between inside and outside of membrane and to regulate and control osmotic pressure.

Note that the actual values in the equation (5.14) are negative, i.e., osmotic negative pressure values, and when the concentration of extracellular ions is high, the water molecules are permeated to the outside of the cell, $(\Delta p)_{osm}$ depending on the ion concentration c of the CH polymers in the cell.

When the concentration c of hydrocarbon polymers ions in cells is high, water molecules permeate into the cells, the space occupied by the CH polymer ions at the center of the cells is compact, water is close to the cell wall, and the size of the residual space occupied by the water is determined by c, so that the osmotic pressure depends on the concentration c. And the hydrocarbon polymers generate depletion effect resulting in uneven of charge distribution in the cells. Thus K^+ ions are concentrated near the inside of the cell membrane, and the change of $(\Delta p)_{osm}$ is magnified by z +1 times.

Donnan potential $V_{D, mem}$ means: the equilibrium membrane electric potential is reached when all across-membrane ion channels are in the closed state and under the equilibrium for all elemental ions between the two sides of the membrane in the presence of ionized CH polymer within the cell. Each elemental ion maintains its own concentration difference across the membrane. The Nernst potentials respectively correspond to the concentration differences of every elemental ion. When the ion channel is opened, the equilibrium membrane potential $V_{D, mem}$ is just replaced by the Nernst potential of the ion is used as the action potential. The Donnan potential deduced from the Donnan equilibrium is [11]

$$V_{D,mem} = \frac{RT}{F} \ln\left(\frac{z[K^+]_0^{in}}{\left[K^+\right]_0^{out}} + 1 \right) \tag{5.15}$$

Here F is the Faraday constant. $V_{D, mem}$ is the function of the initial concentration of element ion $\left[K^+\right]_0^{in}$ and $\left[K^+\right]_0^{out}$ on the two sides of the cell membrane. The membrane potential points from inside of the membrane to outside of the membrane. It changes with varying concentration K^+.

When osmotic pressure is negative, because membrane has extremely low tension the water molecule channel is in a closed state, but water molecules can still permeate the assembled phospholipid membrane and do into cells. And it is the spontaneous process. Then the intracellular element ion concentration $\left[K^+\right]_0^{in}$ is reduced along with the intracellular volume increasing, and the concentration $\left[K^{+}\right]_{0}^{out}$ of the extracellular K^{+} is increase due to loss water. , And the Donnan electric potential is fallen.

The Nernst potential of K^{+} ion and the Donnan membrane potential are opposite, while the continuously fallen potential tends until the Nernst potential of K^{+} ions and deviates from the Nernst potential of Na^{+} ions. When the membrane potential decreases to the threshold of the dipole that controls the element ion channel, the channel potential shifts to the low energy position and the potential is perpendicular to the original membrane potential. The K^{+} ion channel open, the membrane potential is changed to the Nernst potential of K^{+} ions. The inversion of the potential causes the membrane to depolarize in the local region. This K^{+} potential jump results in "action potential". This action potential causes the depolarization process to propagate on the membrane and complete the depolarization over the whole cell membrane. The experimental results show that the wave pattern of depolarization propagation has the characteristics of solitary wave [12]. The process of opening K^{+} ion channel, K^{+} ion entering the cell, increasing the total number of microscopic states, and increasing the Boltzmann entropy.

The CH polymer ions are different from the elemental ions, and while one CH polymer ion is charged and combined with many element ions to form electroneutral state that is still only one microscopic state, but same these elemental ions are the z microscopic state. The microscopic state is asymmetry. Therefore, although the local concentration K^{+} ions near membrane within cell by the depletion effect of polymers is higher than one outside the membrane, but K^{+} ions can still enter the cell when the K^{+} ion channel is opened. But water molecules enter the cell, the K^{+} ion concentration will fall. The change in osmotic pressure is also a spontaneous process of Boltzmann entropy reduction. Until K^{+} dropping, the Donnan potential was lowered to reopen the K^{+} ion channel: this is the basis for two-state control of the electrically controlled gate based on the membrane potential. This is two opposite regulation processes of K^{+} ion concentration.

The opening or closing of ion channel protein controlled by two-states under the action electric potential also involves the linkage mechanism of Na^{+} and K^{+} ion channels. The ion channels are an electronic control gate, which is equal to a resistance, and the film is equivalent to the capacitance C under the action of membrane potential. The RC circuit of each ion channel under the same membrane potential has Ohm's law $I = g\left(V_{mem} - V_{Nern}\right)$, where g is electro-conduction and I is the ion current. Assuming that the membrane potential in parallel circuits is only related to Na^{+} and K^{+} ions, then there is differential equation of membrane potential as following [12]

$$C\frac{dV_{mem}}{dt} = g_{K^{+}}\left(V_{Nern}^{K^{+}} - V_{mem}\right) + g_{Na^{+}}\left(V_{Nern}^{Na^{+}} - V_{mem}\right) \tag{5.16}$$

The membrane potential is formed by the Donnan equilibrium, and local conduction of an ion channel is only locally changed into Nernst potential, and although depolarization can propagate on the whole membrane, $\partial V_{mem}/\partial x_i$ is still propagated of the local change. It is also very important that the Nernst potential $V_{Nern}^{K^{+}}$ and $V_{Nern}^{Na^{+}}$ are reversed, and that the alternation in the two-state control described below has an influence on the time-sequence change of the membrane potential, so that the change in membrane potential can be regarded as a stabilization [12]

$$V_{mem} = \frac{g_{K^{+}}V_{N}^{K^{+}} + g_{Na^{+}}V_{Nern}^{Na^{+}}}{g_{K^{+}} + g_{Na^{+}}} \tag{5.17}$$

When the Na^{+} ion channel is opened, the conductance $g_{Na^{+}}$ is very large then $g_{Na^{+}} >> g_{K^{+}}$ and has $V_{mem} \approx V_{Nern}^{Na^{+}}$. When $g_{K^{+}} >> g_{Na^{+}}$, there is relation $V_{mem} \approx V_{Nern}^{K^{+}}$. The switch characteristic makes the membrane potential stable. However, K^+ and Na^+ ion channels exhibit dual stable switching characteristics under trigger each other, this allows the concentration level of these two ions between membrane to be kept in a steady state. This can hedge the effect that the concentration difference across the membrane is fallen by the infiltration of water molecules. However, there must be two kinds of ions to construct this kind of double stable state switching circuit.

The Na^+ ions in the seawater of the earth can balance the intracellular charge of CH polymer ions and K^+ ion combination by means of the phospholipid membrane and maintain concentration difference of Na^+ ions between inside and outside the membrane. All this happens with the Boltzmann entropy reduction starting from the mixing process. The osmotic pressure and membrane stress by drainage function of water molecule membrane channel constructs dimorphic control gate. When the osmotic action reduces the potential of ions K^+ to a critical potential, the K^+ ion channel can be opened, the jump of the membrane potential is further generated, action potential is generated, and a local depolarization process is propagated on the membrane. The membrane potential is maintained to be stable by the alternating interaction of two bi-stable gated of ion channels which are formed by the coordination of the Na^+ ions.

Phospholipids act as semi-permeable membranes which do not allow the transport of ions across membranes. Transshipment must pass through the ionic membrane protein channel. The K^+ ion channel is closed under condition that the intracellular K^+ ion concentration is higher than extracellular K^+ ion concentration, while the deviation takes place the molecular motor, K^+ - Ca^{+2} - ATP enzyme will complete K^+ and Ca^{+2} ion passive transport. Therefore, there are two mechanisms and channels for intracellular transport of K^+ ions.

Throughout whole process forming life, cells become “alive” through exchange between inside and outside of the membrane. Then the molecular motor with a class of molecules with passive transport machine related life activity, for example acetylcholine, dopamine, cholesterol, etc., has common characteristic: the transport of the corresponding molecules is accomplished by the spatial configuration change of the receptor molecules by means of matching with the ligand molecules.

The cell with ion channels and molecular motor pathways can be regarded as a mass flow 'heat engine' with two ion sources and bidirectional ion current. They are the basis for "live" movement of the cells.

5.4. Folding, Hydrogen Bonding, Structure and Function

Folding is the spatial structure generated between residues by means of hydrogen bonds or Van der Waals forces on the basis of the primary structure sequence of the protein molecule, and comprises secondary, tertiary and quaternary structures. This folding may spatially construct ligands or receptors on the protein molecule that confer specificity and stability to the life activity. Folding is defined, with high fidelity in the genetic process; they are therefore called the "second genetic code" of life. And is a problem that has not been fully revealed in the central law. Although we know the primary structural sequences of 17 ten thousand proteins, only 1.2 ten thousand spatial structures are known. And only 1000 are really different. How to predict the spatial structure of a protein requires knowledge of the rules for its folding. It is only intended here to discuss the spontaneity of this folding process, explaining the progression of proteins from folding to "live" life. It is likely that this will help establish some thermodynamic guidelines for the folding process.

5.4.1. Hydrogen Bonds

Hydrogen bonds are intermolecular covalent bonds, and are H-Y partial covalent bonds formed by Coulomb electrostatic interactions between H core in polar hydrogen bond donors, resulting from charge transfer from covalent bonds in X-H molecules or in side groups of molecules, and Y atoms or ions having an electron rich pair or in side groups of molecules. Hydrogen bonds may be formed either intra-molecularly or inter-molecularly. Molecules in cells have a large number of H atoms, and the participation of hydrogen bonds in the primary structure and constituting the secondary and higher order structure of the molecules of the living matter cannot be avoided. Chinese scientists have imaged intermolecular hydrogen bonding [13], confirming that the hydrogen bonding is covalent. This is similar to Boltzmann entropy reduction in the quantum entangled state of the electron spin as the H molecule is generated, and hydrogen bonding is also a spontaneous process (not discussed here again). But the hydrogen bonding energy is much smaller than the covalent bonding energy of the molecules so much less energy is generated or consumed in the formation or deconstruction of hydrogen bonds. But also provides the possibility for the reversible behavior of life and the generation and the presentation of various functions of various structural forms. Hydrogen bonding has naturally a very important role in the life formation process.

5.4.2. HP Folding Model

The HP folding model refers to the problem of spatial ordering of hydrophobic groups H and hydrophilic groups P during polymerization of molecules to generate protein-level structures. The core of this problem is how the hydrophobic and hydrophilic groups are ordered so that the most hydrogen bonding problems occur. For example, the PHPPHP sheet with 6 amino acids has the largest number of hydrogen bonds capable of being generated in a portal space configuration. Thus, it seems to be a spatial sequence problem, in fact, a problem of maximizing the number of hydrogen bonds formed in space by molecular sequences; i.e., the problem of minimizing the electron number of spin microscopic states. An equivalent problem is that the hydrophobic group has the least potential to form hydrogen bonds. Linearity (directionality) due to hydrogen bond formation; the H with hydrophobic group must be on the same line with the -C-C-chain; under these two conditions, the hydrophobic group H cannot be present in the angular position of the fold, and for two H atoms in the peptide bond group on the angular position can still have 2 directions in which hydrogen bonds are formed. And two hydrophobic groups H on the two straight line section of the peptide chain are taken at two opposite positions. The hydrophobic group could little lose a position or an azimuth to form hydrogen bonds. The four amino acid residues folded according to the β beta turn, and the two amino acids in the sequence on 1 and 4 positions must be hydrophilic groups to form hydrogen bonds to ensure that the turn fold is stable. This results in a folding of PHPPPPHP which is more stable than PHPPHP. The CH polymer must therefore take a folded form in an environment where hydrogen bonds can form; rather than the spatial morphology of the molecules produced by the random walk model. The protein folding process is a spontaneous Boltzmann entropy reduction process. The folded protein spatial configuration will also take the most stable structure. The HP model is also a model for the arrangement of the sequences of the hydrophobic residues with the least hydrogen bond in the CH polymer. Thus, more folding is the steric folding configuration of the CH polymer corresponding to the primary structure of the protein in an environment that can form hydrogen bonds (including dissolution in water) with minimizes Boltzmann entropy.

The HP model was concluded to be 1. The amino acid sequence of the protein is a prerequisite for the formation of a stable folded structure. This is the basis for the framework Model (Faework Model) of Anfinsen, the essence of this model is a primary information sequence model. 2. The collapse of the hydrophobic during folding is shown by taking the para-position spatial arrangement of H and H after folding.

5.4.3. Non-Ergodic Folded

When a hydrophilic group is introduced into the hydrophobic group of the CH polymer itself, the CH polymer can become a system compatible with water and aqueous solutions. In this system, the above-mentioned folding of the CH polymer having a hydrophilic group and a hydrophobic group takes place. But also, is the spontaneous process that must occur. Necessarily, the polymer molecules are polymerized into a sequential-level structure according to the formation rule of the HP model. The secondary structure of the CH polymer cannot be randomly oriented again when first level order structure is defined. Why is this folding not an ergodic process but rather an extremely rapid process? First, going through all possible spatial configurations in time and space violates Boltzmann entropy reduction spatiotemporally. Many of them will be the process of increasing Boltzmann entropy. Folding has the effect of reducing the Boltzmann entropy of the molecule, and spontaneous process generation is naturally unlikely to be ergodic. The tropism of folding also makes it impossible for folding to occur with increased Boltzmann entropy. Random roaming folding is not possible. Second, HP folding is "defined action" or not "discretionary action". Third, folding is the reverse of the baker's transformation of pulling the dough into noodles. The folding of the long-chain CH polymer can be accomplished with a very limited number of inverse baker's changes and with a "defined action". The folding process of protein molecules cannot therefore be described using ergodic models.

How can one get a CH polymer with a defined sequence? Ziegler-Natta bicomponent coordination initiators in stereo-structured polymerization have been used in the artificial synthesis of stereoregular polymers. A range of sterically regular polymers from C2 to C5 was produced. This indicates that the coordination of some certain ligands and receptors is such that the steric configuration of the CH polymer is regular. While polycondensation is considered to be a coordination reaction of ligands and acceptors between lateral groups of the CH compound, when byproducts such as water, alcohol, or ammonia molecules are produced, the CH polymer formed may be stereoregular. Polycondensation copolymers are produced when two different by-product small molecules are produced simultaneously. Thus, the stereoregular and arranged-sequence polymers required to form HP folds under natural conditions can be a spontaneous process. The polycondensation processes generally is spontaneous processes in which Boltzmann entropy decreases that have been discussed in section 5.1. The peptide bond of amino acid molecule to form protein and the phospholipid bond of RNA and DNA to form DNA are all undergone the process of polycondensation reaction, and the RNA and DNA are all made up by using DNA as template, and the peptide chain of protein is transcribed according to mRNA template, so that it can ensure the space regularity of protein primary sequence.

Thus, the certainty of folding derives from the certainty in protein-level structure and the folding rules that the HP-like polymers form hydrogen bonds to a maximum. This determines that folding is the predominant form of life material.

For example, hydrophobic groups surround both water molecule channels and ion channel proteins. This may be considered a recurring ring of HP folding. This allows the hydrophobic group of the channel protein molecule to be compatible with the hydrophobic layer in the middle of the phospholipid bilayer membrane. Changing the $2 \times n$ residual microscopic state incapable of forming hydrogen bonds in the HP model into a mixing state with the hydrophobic group, wherein the Boltzmann entropy of the mixing process, namely the number of microscopic states corresponding to the number of molecules, is also reduced. The inlay process of channel proteins on cell membranes is therefore also spontaneous.

The problem of folding is discussed in terms of the spatiotemporal view of random walk and ergodic states, because there has previously been no concept of electron spin potential $\langle TS_B \rangle_e$ having the ability to reconstruct electron momentum space configurations. Folding is simply a rapid process by which all residues assemble the reconstruction ability of $\langle TS_B \rangle_e$ that correspond all electron spin momentum. All residues will pool everybody's efforts and the folding will be completed by high speed reconstruction. The timescale required for such a reconstruction should also be bring into corresponding with the timescale of the momentum variations in the corresponding distance space $\langle p\not{V} \rangle_e$. The folding that actually occurs also does not support ergodic.

5.4.4. Function of the Folding Structure

The original protein is naturally produced by elements, but has no specificity and stability necessary for it to have some certain function of a living body. This function uniquely determines that it is procedural in which screening for primary amino acid sequences and for various folds is also included. The folding at this process may be a compatible with non-specific processes such as adhesion mechanisms. Reference may be made to a bonding Model (Diffusion-fusion-addition Model).

However, the folding process in living organisms requires a high degree of specificity, i.e., a high degree of certainty, does not allow incorrect folding, just like a gene sequence goes wrong and is likewise pathogenic. (e.g., mad cow disease).

In particular, it is noted that transcription of mRNA is carried out in the cytoplasm when protein growth on ribosomes, i.e., in mixed solutions of a wide variety of protein. There is a unique ligand to the protein to be synthesized, and therefore, when an amino acid is incorporated into the protein by the triplet confirmed by tRNA as a ligand, the unique ligand should be binding capable to the corresponding position by hydrogen bond or by secondary bond as long as it has a landing site. Subsequent growth of the primary sequence, while folding is completed. The way in which folding is accomplished on the ligand can be referred to as the "template folding mode". The folding of this Growth Model starts at a specific site, consistent with the notion that synchronous growth is highly specific in the Model (Nuclear-coding-Growth Model). It is also consistent with the view of high-speed processes when splicing amino acid is folded at multiple ends in an Imposition Model (Jig-Saw Puzzle Model). We supplement the concept of "template folding "for these models using ligand molecules as templates.

This ligand template helps complete folding and also completes protein sequence growth. Binding of a ligand to a receptor forms hydrogen bonds or other secondary bonds, and is spontaneous in that Boltzmann entropy decreases as forming covalent bond. When the ligand and receptor protein molecules are separated from the ribosome, the energy of the ligand from the protein after folding is completed is equal to the energy released when hydrogen bonds are formed by folding the spatial configuration. This does not require additional energy expenditure. Furthermore, hydrogen bonds are formed with net energy expenditure in folding, and the energy released by ligands can be compensated energy removing template at the end of folding in the procedure. The overall folding process can therefore more reduce the number of electron spin microstructures by the formation of more hydrogen bonds. Thus "template folding" can occur spontaneously. It ensures that the folded protein has specific functions.

In the case of no prior determination of the corresponding ligand, even more than one ligand may be present, and we have not been able to predict the spatial configuration, i.e., the "second genetic code" after folding, based on thermodynamic criteria directly and exactly, according to the primary sequence of the amino acids in the protein. Without the protein ligand molecules, the domains and structural domains resulting from folding can still not be predicted.

The function of a protein is determined in a sense by the nature of the "acceptor" constructed by the structural space created after these folds. The vital activity depends on the signal generated by the ligand after binding to the receptor and the subsequent molecular activity. It builds a link between different molecules. There is a concept of the living body as a whole. For example, elemental ions can cause the conformation of the channel proteins on the ion pump on the molecular motor to change and the ion to reverse transport. Realize the balance of ions inside and outside the cell membrane and maintain the stability of the membrane electric potential. Various molecules in the cell can normally move. Thus, folding is the key of a protein with some function and become "live". Predicting the second genetic code may require prior confirmation of the function of one protein.

The folding principle can therefore be generalized as: 1. a target protein criterion. The proteins locked by a particular function, look for possible ligands. 2. Protein chain minimum space configuration number criterion. Through hydrogen bonds or secondary bonds, the number of -C-C- bond rotatable spatial configuration is minimized or goes to zero. 3. The atomic bulk density is maximized. The minimizing protein space volume can be ensured.

The "void" concept as a receptor at the molecular level is also evolving process during the evolution of life. In the process of forming a multicellular organism, when the same kind of cells are accumulated into an organ, an organ membrane is formed outside the organ, and the cell accumulation also forms some voids in the organ; the liquid phase composition maintains the balance between the inside and outside of the cell by means of cell membrane and controls the passage of all the 'material' on the cell membrane. When animals are evolved, a 'void' is formed between the membrane of each organ and the skin that separates the environment from life body. Chinese medicine is well aware of that there is a 'meridian' system constructed by the gap between the skin and organs. It is connected with each organ and called as "Qi". Although the clinical therapeutic effect of acupuncture on the meridians cannot be denied, it is still under question because no anatomical structures can be found. But imagine how the "receptor" spatial structure could be found on the straight-chain molecules after opening the protein' folding? Similarly, when you separate the organs, you go where to look for "meridians". It should be originally a "void" between organs. The "needles" may be ligands to these receptors as "acupuncture points".

The "void" spaces of the meridians are also composed of liquid phases, and the body fluids are in contact with the whole human body. Surface tension and osmotic pressure can drive such body fluid flowing outside the organs and maintain extramembrane transport and equilibrium between organs. Moreover, the movement of the lungs is analogous to the driving action of the diaphragm pump, and the body fluid system should have an overall flow. This should be the unique "Qi" system in TCM. It may be a system of "void" space construction between organs. A system parallels blood circulation. This means that various "spaces" in a living body are also saturated with "information" and functionality.

5.4.5. α Spiral and DNA Double Helix

Another spatial configuration of proteins is a α helix. For fixed pitch, 3.6 residues rotate to 360^0. Residues on α helix at this distance form hydrogen bonds. The peptide chains are made to have a deterministic spatial configuration, rather than a random morphology. Therefore, the α helix not only has space certainty, but also greatly reduces the number of space configurations generated by C-C chain rotation of protein molecules. But also, a process in which Boltzmann entropy spontaneously decreases because of the formation of hydrogen bonds. The α helix structure of the protein thus produced is spontaneous.

The double-helix structure of DNA is a landmark finding in the development of life science. Hydrogen bonds can be formed between all purines and pyrimidines in the two phosphodiester chains. Among them, two hydrogen bonds are formed between A-T and three hydrogen bonds are formed between G-C. Since both nucleotides and bases are cyclic molecules with planar structure, the 10 phosphodiester bonds rotate 360^0. Except Boltzmann entropy reduction due to hydrogen bonding; and the hydrophobic groups on A, T, G, C would form a built-in arrangement. The hydrophobic collapse effect is also a process in which Boltzmann entropy spontaneously decreases. DNA duplex formation is spontaneous.

Energy is consumed to unwind the double helix structure. The specialized enzyme precedes the receptor-ligand coordination, which should generate energy that provides the initial energy to open such double-stranded structure.

The assembly of structures may be a collection of many cellular structure elements. If it is said that the β sheet is a brick-like member of the protein space structure and the α helix is like a rope, and further bundling assembly can form some rod organelles needed in the life body, such as muscle fiber and the like.

The tertiary structure of a protein molecule is either built up by secondary bonds or by spatial domains (domains) formed in various combinations of helices and beta sheets. Independent polypeptide chains as subunits are spatially combined into 4-level structures of proteins by means of secondary bonds, such as hemoglobin, insulin, which are a kind of poly-polymer. They can be a di-polymer, a tetramer, a hexamer and a multimer. The spatial configuration and the polymer are important and have specific functions. But the structure is also spontaneous. New needs may motivate new structural functions, but may also be a lengthy screening and evolutionary process.

5.4.6. Recognition and Operon

The function of the protein is specific. Therefore, what are tens of thousands of proteins that can make them play their own roles? Among these are the functions of "recognizing" proteins to each other.

How do receptors and ligands with a special configuration match each other? It appears that this is dependent on random collisions between molecules. The collision frequency depends only on the concentration. Because the temperature and pressure are constant during normal life activities, there are only little fluctuations. But this fluctuation may be the most basic "signal", being the adjustment signal ignored. There is the influence on above matching process. As discussed further below.

Based on the Boltzmann entropy under the Maxwell-Boltzmann momentum distribution, searching the partition function of matching probability of the receptor and the ligand is the basic theoretical framework for the biological physics to discuss the life activity. This theory gives the two-state regulation mechanism between receptor and ligands. However, the matching state and the non-matching state are two kinds of energy states for the receptor itself. So, these two states are not determined by probability. If the probability is determinate by the certain concentration, the premise of the problem becomes how the concentration is regulated. The concentration of receptors and ligands is regulated by another some information or concentration. This is not two-state control itself. Logically, it becomes infinitely traceable.

How well this statistic is understood based on the certainty of a match between receptor and ligand? Consider that in performing a particular life activity, a large number of molecules of the same type need to be mobilized to perform the same defined activity. Therefore, the life activities are statistical, such as in the process of matching receptors and ligands, but the reason that the statistics are logical is only the uncertainty generated when the number of ligands is less than that of receptors. That is, when the ligand molecule is mobilized to participate in the process, the randomness is likely to occur with receptor-to-ligand certainty if the number of ligands is less than the number of receptors. Such as receptor-ligand problems on various transport channels on the cell membrane. But, when the ligand concentration is sufficiently large, there is no randomness, and statistical theory needs to be improved.

The regulation problem of lactose operon in two-state model, whether positive regulation of recruiting of RNA enzyme to settle in the promoter region mediated by activator proteins(AP), or negative regulation of the repressor proteins at the CAP gene junction in front of the mRNA promoter, and the corresponding activation protein and repressor are both ligands. Both of them are the receptor-ligand regulation problems for only identified region on mRNA as receptor, and the corresponding protein are as a ligand. Both matching problems are all two-state controlled according to existing experiments, depending on the concentration of the activator protein and the repressor protein. Above a certain threshold concentration, the probability of recruiting corresponding protein is constant, i.e., independent of the concentration of related protein. The probability of this pairing is directly proportional to the matching energy value $e^{-\Delta\varepsilon_{ad}/k_BT}$ and is only related to the bond energy release of the activator protein matching to the RNAse. It is therefore still a very well-defined match between the two that can initiate the transcription process of the mRNA on the DNA. In fact, is not related to the concentration of transcription factors.

This P_{bound} number of receptor-ligand matching is the probability that the activator protein matches with the RNA on promoter at mRNA and RNAP matches with promoter respectively, that is, the total probability of synthesis about activator matching with RNAP. The number of activator molecules A and the number N_{NS} of non-specific promoters and the ligand number P are related to the probability of such receptor-ligand binding. The energy released when they bind to the mRNA and when the activator protein forms a bond with RNAP is the $\Delta\varepsilon_{ad}$, $\Delta\varepsilon_{pd}$, $\Delta\varepsilon_{ap}$ respectively. It is contemplated that both proteins are in place on adjacent regions of the mRNA chain. The probability that the two proteins adhere to each other is fixed. Above the same critical concentration A_{CRI} of the activator protein, the function is essentially unchanged and F_{reg} is a regulatory factor [14]

$$
\begin{aligned}
P_{bound}\left(P,A,N_{NS}\right) &= \frac{1}{1+\left(N_{NS}\,/\,PF_{reg}\left(A\right)\right)e^{\Delta\varepsilon_{pd}/k_BT}} \\
F_{reg}\left(A\right) &= \frac{1+\left(A/N_{NS}\right)e^{-\Delta\varepsilon_{ad}/k_BT}e^{-\Delta\varepsilon_{ap}/k_BT}}{1+\left(A/N_{NS}\right)e^{-\Delta\varepsilon_{ad}/k_BT}}
\end{aligned}
\qquad (5.18)
$$

The derivative of P_{bound} to A does not vary with the change of $\Delta\varepsilon_{ap}$ value, P_{bound} and $\Delta\varepsilon_{ap}$ the corresponding single-value functional relation shows that the P_{bound} original function form is not important.

There is also a probability of binding between the activator protein and RNAP in solution. The probability of matching between two protein molecules to each other is only 100%, after the binding of the receptor ligand for the promoter and the upper region has been individually completed on the DNA. This recruitment mechanism is therefore itself a deterministic regulation. And has no statistical significance. The only probability that can occur in this association is the change in the number of hydrogen bonds or secondary bonds with respect to the $\Delta\varepsilon_{ap}$ value, i.e., the number of electron spin quantum microscopic states. The following relationship can therefore be used to estimate the number of hydrogen bonds formed in this receptor-ligand binding. According to experiments $\Delta\varepsilon_{ad} = --13.12k_BT$, $\Delta\varepsilon_{pd} = -5.3k_BT$ and $\Delta\varepsilon_{ap}\,/\,k_BT$ = -4 calculations, from

$$\Delta S_B = k_B \ln \frac{1}{P_{bound}} = k_B \ln \frac{W_{rec-lig}}{W_{rec} + W_{lig}}$$

$$P_{bound} = e^{-\Delta\varepsilon_{ad}/k_B T} \tag{5.19}$$

$$\frac{W_{rec-lig}}{W_{rec} + W_{lig}} = 0.0183 = \frac{1}{54.6} \quad \left(for \quad \frac{\Delta\varepsilon_{ad}}{k_B T} = -4 \right)$$

Where the subscripts *rec* and *lig* are the binding probability of the activator protein in the promoter upper region and the probability of RNAP binding to DNA, respectively. As described above, they are constant after the solute is greater than a critical concentration. The actual effect is to the adhesion between the activator protein and RNAP. This adhesion means that almost 118 pairs of hydrogen bonds are formed between the activator protein molecule and the RNAP molecule. This reflects the extremely precise receptor-ligand match among these. But also release considerable energy to supply the energy required to initiate protein synthesis.

Similar repressors regulation may be accomplished by two kinds of repressor proteins using bi-stable state mode with negative feedback. Both repressor protein molecules, acting as ligands, complete the matching process to the promoter receptors on the mRNA. The magnitude of this binding energy is also reflected in the number of hydrogen bonds or secondary bonds that are generated when they bind to the promoter on the mRNA. The difference with the activator protein is that the number of first repressor molecules in the cytosol affects the probability of binding on the mRNA, but the second kind of repressor molecule is produced after the first repressor molecule completing the receptor-ligand matching. Once binding is complete, the amount of second repressor molecule produced is determined, as is the amount of first repressor molecule produced by the feedback around the promoter. The bi-stable control precludes matching randomization of the first repressor molecule as ligand. This repression process is deterministic. This receptor-ligand matching is also a process of forming hydrogen bonds or secondary bonds, and the transcriptional process, including regulation of transcription, is a spontaneous process.

The signal pathways inside and outside the cell membrane is a series of specific biochemical processes in which cytokines are activated. Generally, cholesterol lipid molecules pass through a cell membrane to be jointed with a target receptor, or polypeptide is combined with the receptor outside the membrane to change the space configuration of trans-membrane protein, so that signals are transmitted from the domain outside the membrane to the domain inside membrane, and then the domain is matched with a next-level ligand, and the next-level pathway is activated through modification such as phosphorylation. A specific biochemical process is initiated. That is, the series of receptor-ligand sequences are activated by cytokines. It can be considered as a process of randomness at startup to turn to specificity determinism.

The specificity of life processes has advanced the molecular structure involved in life processes to become more and more refined. The "second genetic code" in the spatial configuration of the particular protein folded and helix configuration defines the specific "function" for a protein. The regulation mechanism of the receptor-ligand combination makes all proteins to have some orderly functions. The spatial momentum sequence evolution produced by the kinetic process may become a time series of processes. The "biological clock" in the life rhythm is a periodic solution of a system of kinetic equations. Most processes of life activities are such automatically completed "self-care" processes. No additional stress regulation needs to be initiated.

When the original activating molecules for starting a specific process are continuously traced, too many contradictions between specificity and the cooperative requirement of the whole regulation network also occur. The regulation of the whole life cannot be realized by a plurality of regulation 'knob'. The signal that can be perceived simultaneously throughout life can only be a very simple signal. Such signals are only likely to be the most fundamental variables in a thermodynamic system, such as temperature and pressure. The abnormality of small molecule concentration such as O_2, CO_2, or concentration of element ions such as H^+, Na^+ , K^+ , and Ca^{+2} in blood, lymph, and inter-membrane body fluid in a living body should start up some process of regulating the vital activities of some organs. This is the basis for the treatment of disease that physicians are still using today. The Nobel Prize of biological and medical in 2019 awards the discovery of the perception of hypoxia to regulate multi-physiological regulation processes of human bodies. This is also the meaning of the pathway, i.e., through few initiating signals, many specific biochemical and physiological regulation processes can be accomplished.

Regarding the life forms as thermodynamic equilibrium systems in the respective spaces surrounded by membranes, there is a deterministic distinction between the membrane spaces and the domains, and the abnormal fluctuations are probably the most primitive initiating signals.

5.5. Endocytosis, "Feeding" Function and Evolvement

Cells with membranes, which are truly "viable", must have the functions of energy supplementation and metabolite excretion. The molecular structure of transport membrane proteins for glucose molecular has been clarified by the progress of modern biology. The protein structure obtained by Ning Yan, a Chinese scientist in 2013 is a terminal result of the research on the food function of a living body. This has found the "entry" for a series of previously available processes in which glucose is subjected to a series of enzymatic chemistries in the membrane of cell to yield ATP that is life activity "energy currency".

However, it is necessary that such delicate glucose transporter proteins are not prepared in advance and mounted on cell membranes before life tracing is "alive" at a distant age. The original "eating" must be accomplished by a spontaneous process. For organisms that have just had membranes, the mechanism by which communication can occur outside the membrane is only possible by "endocytosis" function of the membrane.

What is the original natural force of the membrane to accomplish this endocytic action? An original body with a membrane, in water, may encounter a variety of particles smaller than its own dimensions, impact or adhesion of hydrophobic organic particles, solid particles, which may or may not wet the membrane. Their adhesion action W_V to the membrane is characterized by surface tension formed after they are in contact with the membrane and before the difference between the surface tension of the water and hydrophobic liquid and the surface tension for the water and the membrane. Assuming a hydrophobic micro particle with radius $r << R$, the cell membrane radius R ; the hydrophobic particles graze in the cell membrane in random motion with a small pitting with radius of curvature r. Then radius of the cell membrane behind the inflection point where the radius is equal to r_1 from the concave to convex transition becomes gradually as R.

$$W_V = -\Delta G = \sigma_{W-M} + \sigma_{W-H} - \sigma_{H-M} \tag{5.20}$$

Where the subscript W is the aqueous medium, M is the membrane and H refers to the hydrophobic material. The contact point of the hydrophobic particles and the membrane has a tangent angle defined as the contact angle θ of the hydrophobic substance and the membrane, and the equilibrium relationship between the surface tension has $\cos\theta = \left(\sigma_{H-M} - \sigma_{W-M}\right)/\sigma_{W-M}$, Thus it is provided with

$$W_v = \sigma_{W-H}\left(1 - \cos\theta\right) \tag{5.21}$$

If θ = 0, it is completely hydrophobic, and the hydrophobic substance cannot wet the membrane; that is to say $r = r_1$. But, generally not θ = 0, since the force of grazing produces the contact angle. So, it will always be $\theta \neq 0$.

Therefore, endocytosis occurs even when the membrane is impacted by an organic substance that is completely non-wetting. There is tension in the convex curve of the cell membrane. The surface tension has a resultant force in the radial direction. This resultant force is opposed by an additional pressure differential Δp across the membrane, and the pressure on both sides of the cell membrane is unequal. That is, there is no concentration difference between the inside and outside of the membrane, and the pressure inside the cell is higher than the pressure outside the membrane due to the curvature of the curved surface alone. Therefore, the osmotic negative pressure must first balance the additional pressure created by the curved surface. Whereas the pressure in the cells localized at the concavities will be lower than the extracellular pressure. This additional pressure is given by the Young-Laplace equation

$$\begin{aligned} \Delta p &= \sigma\left(\frac{1}{R_1}+\frac{1}{R_2}\right) \\ \Delta p &= \frac{2\sigma}{R} \end{aligned} \tag{5.22}$$

However, according to this equation, the change in the radius of curvature of the film due to the crater occurring due to the collision and the change in the radius of curvature of the film in the vicinity has the following relationship

$$\begin{aligned} &R > r_1 >> r \\ &|\Delta p_r| > \Delta p_{r_1} > \Delta p_R \end{aligned} \tag{5.23}$$

The additional pressure in the pocket is negative and the radius of curvature becomes positive after passing the point of inflection, here radius is r_1, from the smaller radius r of curvature of the pocket. The transition section has a radius of curvature $r_1 < R$ that changes from the inflection point back to the original radius of curvature of the cell membrane. The tension and additional pressure at r_1 in the membrane of the convex curved surface are greater than the tension and additional pressure on the cell membrane without deformation. The membrane at r_1 will become more prominent. Due to $r_1 > r$, the maximum negative additional pressure in the cavity in the membrane will deepen the cavity. And the additional pressure in r and r_1 two places will continue to enlarge the deformation of the pitting until it encloses the endocytosed material until the object matter particle to be "eaten" into the cell membrane in an endocytic manner. This process is a surface free energy reduction process and is therefore naturally a spontaneous process. And this process results from random collisions with the cell membrane. It is a completely autonomous process that starts automatically. The nature of the endocytosed particles may be any material, provided that the particles have sufficient momentum to cause collision pits on the membrane, and an endocytosis will occur.

This process does not require the generation of new membrane materials, does not generate the need for additional energy, and is a completely spontaneous process.

When intracellular metabolites are excreted to the outside of the cell, the metabolites are small molecules and are accumulated in the inner side of membrane due to the depletion force, and the local protrusions are more aggregated to form vesicles. The vesicle can be used for forming a spherical bubble with a minimum area and accommodating the most metabolites to be discharged outside. This "exocytosis" is also a spontaneous process. Some of the membrane surface lecithin is lost during endocytosis. However, vesicle size is less than 0.01 relative to membrane surface area, which is assembled, and this loss can be compensated for by the compactness of the array.

Endocytosis and exocytosis may be therefore the primary metabolic processes of the primordial cell. These cells also function in higher organisms, as macrophages in the immune system phagocytose foreign invaders in this way. Virus invasion is the passive swallowing of virus into normal cells and the spitting of proliferating virus. Mitochondria in cells are likely to be endocytosed in the history of eukaryotic cell formation. Recent outbreaks of new coronaviruses in China are highly contagious. The new 2019-nCoV by means of S protein on viral spike matching with ACE2 dimer on membrane generated invagination and resulting in endocytosis go to cell.

5.5.1. ATP

The cells have the function of "eating" or in order to harvest energy, the earliest energy that can be acquired may be the simplest possible ATP. Although we have found that organisms can efficiently obtain ATP and NADPH, which are energy units of organisms, from carbohydrates and even cellulose through a series of enzymatic chemical reaction processes. But these two energy units are just the simplest energy molecules and may be also the most likely molecules generated by spontaneous processes under natural conditions. And are molecules that can be utilized when cells become "living" in the early stages. And then a refined digestive system is formed finally in the evolution process to complete the chemical treatment process of all ingested natural substances. Therefore, the problem needs to be simply reviewed for APT molecules.

The molecule is a ribose molecule in an adenine molecule that is phosphorylated by three phosphate groups. ATP is identical to a unit on the phosphodiester chain of DNA. ATP should appear earlier during life evolution. This energy unit is converted to ATP by the action of inorganic phosphate Pi and diphosphate ribose + Adenine (ADP) upon irradiation with photons. ATP hydrolysis reversibly returns to ADP + Pi. The reversible reaction process can also be used for driving an ion pump on a membrane protein constructed by ATP enzyme protein by consuming the energy for hydrolyzing ATP, and transporting ions across a membrane by a reverse ion gradient, such as a Na^+ (K^+) ion pump, so that three Na^+ are pumped out of cells and $2K^+$ are pumped in at the same time. ADP + Pi produced by ATP decomposition is converted to ATP by other forms of energy, such as H^+ ion pump transport into membrane, and the conversion of ADP back to ATP is carried out at ATPase.

It is also worth noting that the energy of ATP reaction is 20kBT, which is close to the energy level of hydrogen bond. This is related to all receptor ligand binding and separation processes in life, which are accomplished by the formation and dissociation of hydrogen bonds. The processes are both deterministic and reversible, so that the overall energy consumption level is minimized. In addition to the synthesis of proteins, life activities are also largely involved in the formation and separation of molecular covalent bonds. ATP is therefore matched to most of the vital activity. Including on the molecular motor proteins that transport Ca^{+2} and K^{+} ions, there is also a molecular "foot" of ATP.

5.5.2. Evolution

The viruses only with RNA can enter cells by endocytosis. Many cellular organelles, including organisms with DNA molecules, could also be gradually introduced into cells by endocytosis. Because there is no inherent logic necessary to evolve both DNA and RNA molecules simultaneously within the cell, so that they can each be produced spontaneously in nature. Membrane endocytosis plays a very important role in the convergence of life matters during long evolution process. No matter what is useful or useless, the user can be screening out naturally after collecting the useful matters.

Chapter 6

DISCUSSION AND PROSPECT

Some of the work once addressed in the continuous media field was discussed above. It is proved that the pressure generated based on elastic collisions in an ideal gas is a fundamental force for generating gases congregation state. It corresponds to the translation momentum of gas molecules or atoms. Based on Maxwell's demon problem, the concept that momentum distinguishable particles are also spatial distinguishable is proposed. It is applied to Boltzmann statistic, and two concepts of local off-domain sub-statistics and spatial degeneracy states are proposed. The certain Boltzmann entropy value can be obtained under thermodynamic equilibrium condition. The principle of interaction invariance $pV = TS_B$ is equivalent to the Maxwell's thermodynamic relation, and the two space configurations according to the distance space of particle and the size of momentum exist in the momentum field for each kind of momentum in thermodynamic system, and there all are two kinds of potential energy for the same action.

These discussions are limited to the physical category of condensed states. This chapter attempted to raise some questions to see if the potential energy TS_B that could have been previously over looked would also exist in more interactions. To see if this potential energy is compatible with the existing physics. And what I think is interesting questions to discuss. Some already exist, but have not yet been solved. Some arguments have not yet been decided. Take advantage of this book to put toward some ideas. These problems are expected to be interesting.

6.1. ENERGY CONSERVATION AND TIME ARROW

The time translation invariance of energy shall mean the conservation of energy under the same acting force, or the conservation of energy of the same type of momentum, including kinetic energy and two kinds of potential energy. But all thermal effects can be contained in the free energy, including phase transitions, heat of chemical reactions, and heat of mixing, all of which involve different interactions in an isolated system. Thus, free energy does not have the premise of energy conservation. The free energy is not necessarily conserved. It is also important that the system itself is unchanged.

The relationship between phase transformation and spontaneous symmetry breaking is widely discussed in condensed state physics. The spontaneous symmetry break is broadly defined as the invariance of the Lagrange content of a system with spontaneous symmetry break from a kind of certain energy under the action of a transform group, but not to the ground state energy. This system extends all the way to the Higgs' field of a quantum vacuum. There is a spontaneous symmetrical breaking of mass from inexistence to existence, and constituting a sequence of events. The evolution of the universe is a time sequence of events. The sequence of events included the appearance of the original universe from vacuum without any mass under the Higgs mechanism, and the quarks to neutron, proton and electrons arisen from weak interaction and strong interaction, as well as the atoms and molecules under Coulomb force, in which there is electromagnetic field energy of electron. Then under the pressure generated by elastic collision atoms or molecules are gathered into gas, which is then broken by spontaneous symmetry with the changed type of molecular momentum to produce a liquid with rotational momentum and a solid with vibration momentum. There are also various forms of phase change when momentum symmetry breaks down. Until the masses of the fixed star and the planetary system are gathered to the planetary distance space generating in the gravitational action field, but the masses are lost; it is not known whether these dark substances are not visible or whether it's not mass that's missing. From the red shift, Hubble finds the cosmic expansion, and the corresponding energy is called the dark energy, in relation to the momentum of the expansion motion. All of these interactions have corresponding the momentum. And some different particles have all the same canonical invariance. As they have all the co-rotation of the particles. Time is for momentum, and the time scales for different distance spaces are also different. The parallel-existing momentum fields have parallel time-space with different space scale. If you just focus on motion, there's only distance, no space, so there's no continuous time, has only a time gap in sequence. The sets of various distance space are a fractional dimension space and the sets of time gaps in time sequence is also fractional dimension. Also, it is necessary to introduce spontaneous symmetry break-up to see the time sequence of the universe, and the problem can whether the time be flows backwards? Can the earth turn around the sun and back to the origin? It is impossible to enter the same river twice from philosophy. The earth returns to the origin, but the earth is nobody in spring and autumn. The inflated universe should also be irretrievable in the vast ocean. Therefore, we also see what happens when this spontaneous symmetry breaks down? Does time have an origin?

Except for irreversibility due to randomness in deterministic motion, let us don't go back to the starting point. The change between the momentum types by way of the spontaneous symmetric breaking is also no the original space-time.

In order to solve the problem that some bosons have not mass, the concept of introducing quantum vacuum as ground state energy based on spontaneous symmetry broking. The cosmic filled by Higgs field is accompanied by special Higgs' particles as a special kind of commutant of ground state energy, which is the latent vibration energy in the Higgs field (quantum vacuum ground state energy field). Under other interactions, spontaneous symmetric breakdown of energy away from the ground state occurs, producing bosons with mass. The corresponding fermion can also be made to acquire mass. The problem without mass in the Young-Mills theory is solved. The physical concept of the spontaneous symmetry break may be more important because the spontaneous symmetry break in phase changes involve commutation and reconstruction of energy field between different interactions.

Reviewing the phase change process of ideal gas to liquid, the spontaneous symmetry breaking process of which the molecular translation momentum tends to zero, i.e., to achieve the ideal gas ground state energy state. In this ground state, the kinetic energy of translation motion tends to zero, the pressure p resulting from the elastic collision of the molecules in an ideal gas being frozen $(p \to 0)$; instead of the corresponding intermolecular interaction, which are present as van der Waals forces, the rotation momentum of molecule is transferred, the number of molecules of the gas translation momentum being present $N_{g;Tra} \to 0, \quad S_{B'g,Tra} \to 0$

$$
\begin{aligned}
&(p\not{V})_{g\to l} \to 0 \quad \Delta T_{g\to l} = \Delta(p\not{V})_{g\to l} \\
&(TS_B)_{g\to l} \to 0 \quad \Delta(TS_B)_{g\to l} = m\gamma_{g\to l}
\end{aligned}
\tag{6.1}
$$

When this spontaneous symmetry break occurs, the potential energy change of momentum space configuration generated by phase change is the phase change heat. The heat is not only TS_B field energy, but also can be propagated in the field in a vibration mode of thermal radiation, and can also be a heat conduction process by the form of thermal waves in a continuous medium. In fact, this released phase transition heat corresponds to the ground state energy of the frozen momentum.

When both the force and momentum mutation at temperature in the discontinuous phase transition, the energy for the lost momentum is the energy at lowest energy state (such as condensation), and the discontinuous phase transition is a spontaneous symmetric break because the Lagrange's amount is no longer invariant to the lowest energy state.

Supposing that a small closed container in an isolated system contains two-phase with gas-liquid, the temperature is equal to that of the isolated system, and the gas and liquid phase molecular motion is always considered to be in thermodynamic phase equilibrium; the molecules with large momentum enter the gas phase, and the molecules with small displacement momentum enter the liquid phase. However, as described above, when the momentum type is changed in the discontinuous phase transition, that is, the momentum invariance is not maintained, and accordingly, the spatial translation invariance is broken. It is envisaged that the volume of this container may be varied to maintain the same pressure as environment. The condensation process occurs in an insulating environment large enough that the system temperature must be higher than environment for heat release. The same temperature of the evaporation system, also in this adiabatic environment, requires the environment temperature to be higher than this container temperature. Both processes stop when both of the container and the environment temperature is equalized. But these two spontaneous processes directly violate spatial translation invariance. For the container, the boundary conditions are asymmetric. At the discontinuous phase transition temperature, there is no momentum conservation. But this asymmetry is spontaneous direction for condensation. The molecular momentum in the evaporation process is transferred to a high energy state. The symmetry break is not necessarily in the spontaneous direction. Therefore, heat in phase change can reversibly enter or exit the system, which can also be called as symmetrical break, but the spontaneous symmetrical break has a direction pointing to the lowest energy state. It should also be clear where does the lowest energy corresponding to these various momentums reside?

Another form of p disappearance is considered among the interaction forces of attraction, Coulomb force, pressure etc., when there will be at $r \to \infty$, here is $p \to 0$. Of course, the scale of ∞ is completely different for different interaction. So, there is also a renormalized ∞ space. So that from a known cosmic scale there is a ground state space that is a fractional dimension space nested and embedded together which should be a common space. Although this is a vacuum of momentum, the concept of "environment" in thermodynamics still exists based on a "matter".

This space is $\left(p\not{V}\right)_{\not{V}\to\infty}\to 0$ available for any interaction and then also $\left(TS_B\right)_{\not{V}\to\infty}\to 0$. Since energy is only available for motion. That is $S_B\neq 0$ except for Higgs field, this ground state spatial field has a corresponding ground state energy field $T_{\not{V}\to\infty}\to 0$. This ground state energy should correspond to 0K if the time translation invariance for the energy is still reliable. And the 0K state may be correspond to the end of time. But the background radiation which is uniform everywhere in the universe is not equal to zero, which indicates that the ground state energy in the ground state space may not be 'empty'. That is, it is expected that ground state energy is accumulated in the ground state space in various time gaps. This accumulation points to the opposite direction of the arrow of time. In addition to the irreversibility due to randomness in the deterministic action already discussed, the most important would be the asymmetry about the spontaneity of the symmetric breaking process creating the ground state energy field, and the accumulation and growth of ground state energy due to this spontaneity.

Extending broadly this phase change is a spontaneous symmetry break, so that on the sequence of interaction changes, there is an energy sequence of ground state. This sequence is also a time sequence of energy change events. It can therefore be assumed that the spontaneous symmetrical break sequence process corresponds to the time arrow. It is a natural direction of spontaneous processes. From the theory of large explosion, a Higgs field is a field with only energy and no mass, which is equivalent to a field with no momentum, i.e., no $p\not{V}$ field, no S_B field. This field has various possible energies that can correspond to all momentum classes. It can be coupled with any kind of interaction, could commutate between various momentum and interaction. The process of decreasing temperature from the Higgs field without mass, the Higgs particles generated by the spontaneous symmetric self-coupling process, then can complete all subsequent coupling actions. Corresponding to the spontaneous direction, all the series of spontaneous symmetrical breaking processes are also experienced in the micro and macro world with mass. Also, in these processes, the ground state energy field gradually accumulates the generalized phase transition heat generated by various spontaneous symmetric breaks. This energy field is unlikely to annihilate in terms of conservation of energy. Where did they go? They should be at least the ground state energy accumulated in the ground state space. Whether may this be a possible, inconceivable dark energy field? The convention in thermodynamics is to transfer heat to the environment, due to annihilation by the environment. But it has in fact accumulated all the time, the heat of a cup of water is negligible, and the evolution of the planet is only a cup of water to the universe. At least in the Higgs field only a small portion of the energy becomes the world with mass, and the rest of the energy, if it remains in the world, should not be a negligible amount of energy. Does it exchange again for some unknown interaction between the stars that should correspond to the momentum of such expansion? But if the cup accumulates these "waters" with the same age as universe, what is again? This expansion momentum should correspond to an interaction that is not known. This interaction produces a uniform isotropic translational motion of the mass carrier in the flattened universe. The distance space in $(p\not{V})_{Uni}$ is the scale of the universe, but not the scale of nebula. Planetary gravitational interaction is only a local field therein. Dark energy of a mass-free carrier may be a continuously accumulating energy field; but the increasing energy fields continue to increase. Whether or may not the ground state space be regarded also as a corresponding $(TS_B)_{Uni}$ to exist for some interaction? That is, the "empty" of the universe is not empty. And whether the orientation of the cosmic expansion is also the direction of the arrow of time? The corresponding time scale is also the longest, yet unpredictable scale.

When we look back a system moving at the speed of light, when all the speeds of motion in this system are at the speed of light for $\gamma = 1/\sqrt{1-u^2/c^2} \to \infty$, there only are the field. The physical meaning of the Lorentz transformation should be the temporal and spatial scales $t = \gamma T_0 \to \infty$ and $l = L_0/\gamma \to 0$ that is a field seen from the absolute stationary space-time. The inertia system with light velocity motion is without space concept. Time also becomes meaningless. This is because there is no relative motion in this inertial system. For $v' = (v - V)/(1 - vV/c^2)$, always has $v' = c$. When the $V = c$, $v' = (v - c)/[(c - v)/c] = -c$, And when $v = 0$, $v' = -V$. It indicates that there is no relative velocity in this inertial frame moving at the light velocity. There is no relative displacement in the inertia system with the light velocity motion, namely there are not distance, which is the real meaning of the time expansion effect. The time is a concept derived from velocity, and there is also no time without the velocity and distance. Thus, an infinitely time scale in an inertial system with light velocity is pointing for no relative motion. And therefore, there is virtually no time in such a field. In electromagnetic fields, there is no concept of relative velocity, and thus no concept of space and time, i.e., their space and time scales are zero. So, time is with an origin. Although it is now reported that the quantum entanglement speed of light quantum is larger four orders of magnitude than the velocity of light, but is it speed in the mechanical sense? Fields should be different from spaces and fields should be definable as spaces without distances.

Since there is no speed in the system, $E = mc^2$ should be interpreted as meaning that there is only energy and no mass in this system. After the spontaneous symmetry breaking process, some energy in inertial system with the light velocity is transferred to the ground state space, and the inertial system has mass when the energy is lost and the velocity is lower than the light velocity. However, with momentum, there is a concept of space. Both mass and space are thus derived from spontaneous symmetry breaks. Looking back at sequence of these spontaneous symmetry breaking even from an inertial system with the light velocity, these events are one inertial system after another that are far away at different velocities from original point of time. They should just form an arrow of time. Time and energy are coupled together at that is not space. And has time translation symmetry. If does a system not appears to state equal to 0K, the ground state energy might also not be endpoint, or 0K state should be an unreachable endpoint.

The energy field from the beginning of a major explosion serves as the starting point of time, and the inertia system, such as cosmic expansion, which is relatively far away, has an infinite time scale according to the Lorenz transformation. The time scales of all inertial systems with other velocity are all the various time gaps embedded thereon. Would the Cosmic expansion point to the end of the time arrow?

Planetary gravitational interaction is only a local field therein. Dark energy of a mass-free carrier may be a continuously accumulating energy field; but the increasing energy fields continue to increase. Is the ground state space also considered to exist a corresponding $(TS_B)_{Uni}$? That is, the "empty" of the universe is not empty. The huge sidereal mass of the newly discovered early-stage planetary systems in the universe may not be possible to measure deeply without uncovering the mysterious veil in the universe .

This spontaneous symmetry break can be defined by a change of the momentum type. Since this means that the ground state energy can be defined. Spontaneous symmetry breaks can be more physically contained and more generalized. But this does not apply to the coupling effect of commutant. It demonstrates the existence of a potential energy field TS_B for each momentum, but the two momentums that produce the exchange each still maintain conservation of energy and momentum. Although some other break of symmetry may occur during the coupling process.

Finally, it is pointed out that during life formation process the proteins have the aid of hydrogen and secondary bonds and folding and helix structures are spontaneously completed. The "receptor" space formed in the space, although no mass is in it, is indeed a space with information. This information space therefore also should have a ground state space.

6.2. Axiomatic Thermodynamic Law

Is the thermodynamic law established in continuous media, still valid in the quantum world? Just like could the axioms in Euclidean geometry still be applied to non-Euclidean geometries. That is, whether the law of thermodynamics has axiom or not must be proved by at least one form of theorem.

6.2.1. Thermodynamic Law

The first law of thermodynamics establishes the foundation stone of the first industrial revolution, and establishes two basic concepts of work and heat and the un-equivalence between heat and work in the heat engine' cycle, namely the first and second law of thermodynamics in general.

The first law of thermodynamics can be viewed as a form of conservation of energy. In the thermodynamic cyclic process of continuous medium, its statement formula is impossible to have perpetual motion machine of the first kind. After one corrects the "thermal mass" theory, the concept of heat equivalent of work in Joule's experiment show that the first law of thermodynamics is for more general processes.

The impossibility of perpetual motion machine of the second kind was demonstrated by means of the Carnot cycle, for which three declarative expressions about second laws of thermodynamics were proposed as the axiom: Clausius states that it is impossible for heat to flow uncompensated (spontaneously) from a colder object to a hotter object. This is a statement of the highest confidence for the general process; 2. Planck statement is that a system absorbs only heat without releasing heat, and it cannot return to the original state, i.e., a heat engine can have only a single heat source. 3. Kelvin statement is that it is not possible to cool a portion of the object below environment's temperature and extract work. Latter two states for the cyclic process do not the general applicability of any process. The concept of spontaneous processes was first proposed in Clausius statement.

This declarative form for law has very high confidence, so it is axiomatic. But axiom requires mathematical form to make it universal. Work is written as a spatial integration effect of the force. But heat is not derived quantity from the heat equivalent of work. In the rational thermodynamic it is a cumulative function along the heat manifold (which can be mapped to temperature) in the process set. Two important concepts are introduced. 1. The process sets have a time sequence; 2. Heat is accumulated over a time sequence. This concept means that the concept of "heat" is more like the energy than the "work". Work is equal in value with heat, but not equivalent. It is very important that heat does not mean its depreciation in the sense of energy.

The Clausius entropy is a physical quantity defined by the heat and hotness or temperature, which is the physical quantity with directivity. There can be no overestimation of its importance. The entropy function fuses the two laws of thermodynamics together. The set of Maxwell thermodynamic relation equation completes the mathematical deduction proof of thermodynamic axiom, so that they can be applied to any processes in condensed physical. The system in thermodynamics is completely separated from the environment by used of free energy. The concept of heat and work is transformed into energy of the system itself. But it is not conserved. The work of this book is to establish the free energy under the same force. The free energy for the same momentum should be conservational quantity.

Discover of Molecular and atomic led to Maxwell-Boltzmann (M-B) statistics. The statistics based on momentum resolvability gives a way to unify macroscopic certainty with micro-occurrence randomness. In effect, it is a "bridge" of thermodynamics going from the microcosm to the macro matter. The establishment of quantum mechanics is inextricably related to statistics. The solution of Schrodinger equation is interpreted as the spatial probability function of the particle. However, the F-D statistics and the B-E statistics of micro particles are all delocalization sub-statistics, and the concept of spatial divisibility is virtually eliminated. Only the energy level is retained as a concept of "container" in B-E statistics. But it corresponds to the concept of a region. The "interchange degeneracy state" in F-D statistics is equivalent to the concept of spatial degeneracy state. These three statistics are the same because they are not randomly distributed in space based on energy levels or degeneracy states, and can be counted by using a method having domain's momentum divisibility in space and momentum indistinguishable in local space region. Therefore, the basic axiom system of thermodynamics based on thermodynamics statistics can be applied to the world of microscopic particles. Some of the confusing results in quantum mechanics are also more truly interpretable. For example, electron orbit space in atom and molecules, quantum statistics can be regressed to statistics on the probability of a particle for spatial distinguishable within local orbital space. There is certain energy for local random space. The orbit space is one equilibrium state for an electron momentum in atom. But the statistical calculation for all orbits is not that can obtain deterministic results under thermodynamic equilibrium conditions. The statistics of random position for quanta, namely distance space between nucleus and electron, have the certain orbital energy in the statistical sense. This is the certainty that statistics hope to get, and it is also the original intention for the introduction of statistics.

The free energy $G = T + TS_B + p\not{V}$, defined on this statistical basis, is the energy that can be conserved for a particular momentum without transfer of work and heat. The $\Delta S_B < 0$ represents the spontaneous process direction, and the $p\not{V} = TS_B$ equivalence relation of the two potential energies not only clarifies the work and heat equivalent, but also gives the invariance of the acting force and the conservation of the two potential energies for the same kind of momentum. This makes the axiomatic statement of the thermodynamic laws to have mathematical form based on statistics.

Because F-D and B-E statistics are widely used in quantum physics, there is reason to believe that the thermodynamic axioms in the above mathematical forms are applicable, whether for single or multiparticle systems, if only there are spatial statistics of momentum or position.

This force invariance relationship $p\not{V} = TS_B$ is shown in the ideal gas model. But this special case proof can be true for all motion of momentum fields with itself space-time framework? Through the above unified discussion of M-B statistics, B-E statistics, and F-D statistics, it is demonstrated that all statistics under thermodynamic equilibrium conditions are the same in terms of local off-domain statistics with spatial degeneracy. This identity may ensure that $p\not{V} = TS_B$ invariance should be applicable in standard models. But where the boundary of the thermodynamic system can be drawn is problematic. There are concepts of $p\not{V}$ and $\not{V}$ in gravitational fields, but what is TS_B ? It is not known what is p for inflated universe. But there are indeed studies that attempt to describe the universe by used of statistics.

6.3. TS_B IN GRAVITATIONAL FIELD

The gravitational force needs to be discussed specifically because there is no gravitational force in the Young-Mills theory. Generalized relativity replaces gravitation with flexible space-time in mass field, but not without momentum. And does TS_B have in the gravitational field because of the momentum field? How to understand this momentum distribution in the gravitation?

The massive celestial bodies all have gravitational fields and are local spaces in the expansion universe. It is not a fixed and unchanging spatial position in absolute time space; but a distance space formed by the motion of the celestial body and the surrounding star system with attractive force in an ever-expanding universal space. The momentum of the celestial motion with a large mass also constitutes the spatial configuration of the position and momentum of the celestial body. The direction of these celestial bodies along the cosmic expansion is their translational momentum, while the orbit around some star as axis in the local gravitational field is the revolution, which is the isospin momentum of the celestial body; each celestial body also has spin momentum. The celestial bodies may also be charged plasmas or magnets. There are also various kinds of momentum of aggregation and condensed state of elementary particles with their own interaction corresponding to neutron star, red dwarf star and black hole in the universe. That is to say the likely number of combined momentums in the gravitational field is a lot. Perhaps not a simple gravitational force can be fully described. Nor is a kind of TS_B energy under a simple gravitational interaction could fully be contained. In the case of the solar system, which is a familiar planetary system, the planets should not be randomly arranged on the orbit around the sun, the mass of the planet and the rotational momentum of the planet around the sun determine the orbit distance from the sun, and the rotational momentum of the planet and the satellite around the sun determine a momentum space configuration. This is not to be interpreted solely by gravitational forces. The abnormal distribution of the Pluto and the Kuiper belt in the solar system is a phenomenon which is not well explained. As a guess: this may be determined by the process of cooling the momentum space configuration of the dust into planets after the interplanetary dust has accumulated around the sun. The momentum space configuration TS_B potential energy in the local space of a galaxy has never been considered. And according to the force invariability, it should be equal to the gravitational potential energy of the gravitational field, so that it is likely that energy is also absent in the gravitational field and mass (dark matter) is also absent. Or there is only simply a lack of energy. If it is speculated that there is also the Virial theorem in gravitational field similar to that in the Coulomb force field. The kinetic energy is only a quarter of the sum of two potential energies.

There should be a corresponding TS_B for the momentum of the celestial movement in the gravitational field. Since it is only the distribution of the planet that is well seen, but it has not been discussed. Nobody even thinks it's a thing.

We are also looking for graviton. Celestial bodies are carrier of gravity that produces gravity, and that itself is not the graviton. It should be associated with TS_B for celestial system in gravitational field. And it would only be relevant with $\Delta(TS_B)$. Therefore, what force is coupled with gravitation force when the graviton appears? What momentum does exchange with the momentum in gravitational field? The other kind of graviton or may be the exchange between different momentum carried by the star body itself. It may be the generation, collapse or annihilation of the planet. Astronomy studies and observations will confirm the presence of such energy fields. It's just that we may need to look at all observations and discovery reviewed with a new perspective. There will not be only dynamic processes between celestial bodies. If only the attractive force interacts with each other, the universe will not expand. And also has no colorful evolution.

6.4. The TS_B in Quantum Theory

Feynman, Richard says: I think I can say with certainty that no one understands quantum mechanics at present. Physics as understood by Bohr is merely an explanation and not how it naturally works. The reason for this is that the relationship between interaction and motion is discussed only by using the kinetic relationship, and TS_B of quantum physics is lost. The second is how to understand the deterministic and uncertainty problems in quantum physics.

6.4.1. Quantum Uncertainty

The most essential of quantum behavior is the randomness of motion. Uncertainty of the position of electrons in electron orbits cloud of atoms and molecules is discussed in chapter 4. Momentum is also uncertain in the electron cloud. But there are the deterministic orbit radius and energy value by statistics. In conservation of mechanical energy, the position and momentum can be continuously varied in a determined energy level, without the need for discrete momentum and kinetic energy. It is simply not known where the electron is now. Others are deterministic. The quantum motion is random in nature. And it is not changed by the intervention of the measurement. Nor is it generated by the measurement. The uncertainty is the quantum behavior of the particleitself, as long as there is an interaction force within this distance space.

For a single free electron there is a de Broglie wave, this material wave is a classical experiment of position uncertainty in diffraction. This is the result of the intervention of measurement by the crystal lattice. In the double slit experiment, the initial velocity and direction of free electrons are uncertain, but the slit does not interfere with motion of electron and only artificially changes the measurement intentionally. The position distribution of the electrons on the screen when the particles are interfered or in a single slit is observed to be random. That is, there is objective intervention and subjective intervention in the measurement problem.

6.4.2. Wave/Particle Duality

The light has a wave/particle duality different from that of the material wave. Light is a purely electromagnetic field, an energy field. The particle performance needs to complete the energy exchange with the electrons in the metal atoms to be displayed, and is always only an energy field. A material wave, such as an electron, carries two types of momentum and energy by the carrier itself. Free electrons are particles having a translational momentum when they are emitted from an electron gun, but an induced electromagnetic field is present around the charged moving particles. The charged moving particles themselves have wave/particle duality

6.4.3. Double Slit Experiment

The most famous result of this experiment is either non-A, i.e., B, or a superimposed state, i.e., A, i.e., B. There were also Schrodinger's cats who failed to conclude that it was alive and or to death. Double slits do not produce a wave of matter. But are referred to as probability waves and do not have any physical inclusion. However, it was considered that there were two manifestations of the wave and particles after the slits, which should be problematic. The electrons should carry electromagnetic waves in front of the slits. The electromagnetic wave fields may appear at A and B at same time. Just as a light beam splitter can split into two beams. If photons are emitted, the simultaneous presence at A or B is not questionable. The vibration of a single electron on a quantum scale may have a matter wave. Interference fringes of the material wave can be generated. If the electromagnetic waves are still present at the slits, there will be diffraction at the single slit. But only the particle distribution is seen through the single slit. There is interference of the material wave after double slits.

There is only particle property of the electrons in front of the slit that is therefore not correct. The electron for the electron double-slit experiment carries different momentum, translation, electromagnetic field and (corresponding to matter wave) vibration momentum, and can be read as the reconstruction capability of respective TS_B potential energy for three kinds of momentum. It is not needed that three-valued "quantum logic" is used. The electron does have three different momentums. And do the complementarities correspond to lack of real?

6.4.4. Delayed Choice Experiment

The conclusion of this experiment was that the cause and effect relationship was inverted. The original waveform duality of the light is determined by the result to be the initial state. The setting of this initial condition is also wrong. Since light is only electromagnetic, it needs some condition that light shows particle character, e.g., photons excite electrons. There aren't other conditions from wave changed to particulate in this experiment. There should be also no the law of causality.

6.4.5. EPR Effect

This is a quantum state superposition problem. It is generally not considered what effect there is entanglement between the superposition state and ground states of the light. But why white light is often seen, there is certainly the reason behind when some colors are seen. For example, a prism is used to split the light. When ground states of different frequencies are superimposed there should also be a spontaneous process of $\Delta S_B < 0$ resulting from the superposition of quantum state. The entanglement state should be more stable.

They are still electromagnetic waves and the entanglement state can also be subdivided into multiple beams. The parallel superposition states are still electromagnetic fields without mass. Not corresponding to a parallel universe with mass. Information that increases or decreases in the per-bundle local entanglement state, such as information on a carrier wave, can naturally still be transferred in the universe like electromagnetic waves and received in all other local areas. This does not seem to alter the known basic properties of electromagnetic waves. Since it need the condition generating resonance in any location to receive such information. So that it is not always telepathy.

6.4.6. Eigen Wave

When the particles are subjected to $\Delta S_B < 0$ spontaneous process, the released energy $T\Delta S_B$ should generate a corresponding potential $-\Delta T$ and a reaction $-\Delta p$ according to the principle of force invariance, so that the intrinsic vibration of the particles and the corresponding wave are generated. This is the reason which Schrodinger's equation should be a wave equation, not just a form. The complementarity principle has physical basis and not merely an interpretation for natural processes.

Without the uncertainty in quantum mechanics, there is no certainty of momentum that is statistically counted in the distance space of quantum scale. The states in quantum time-space are thus countable. The TS_B potential should just be that potential concept from Aristotle is borrowed by Heisenberg, Werner. But this is true energy. States are thus countable in quantum time-space. But this is true.

6.5. Unification of Force or Heat?

In the history of the overview, Newton extended the contact force between objects to the force of gravity, the force of non-contact between bodies. This has led to the rapid development of physical theory. Analytical mechanics ultimately culminates in all mechanical problems. The Einstein's relativistic theory is that, in essence, the dynamics can also be discussed without the force concept. In quantum physics, the Youn-Mills theory proves the gauge invariance of three interactions, namely electromagnetic force, strong interaction and weak interaction, by using a symmetry method. These methods illustrate that our concept of not requiring force can also discuss the matter movement problem. But why do you go back to search for large unity theory of force? Can this change the fundamental physical fact that interaction changes the magnitude and direction of momentum? Is the chord vibration around the back of the concept of energy, replacing the three different basic moments known with one momentum, whether is it unity force or unity energy? If the forces are unity in the energy sense, that is, the thermodynamic symmetry relationship of Maxwell, eventually they may remain in a force invariance relationship $p\mathcal{V} = TS_B$. This implies the time translational invariance and the space translational invariance.

The heat directly related with TS_B is a physical quantity that has not been well described. We have only advanced one step from "heat mass", which is the energy exchanges forms between system and environment. However, this is not compatible with the mixing heat, reaction heat, and phase change heat in the system. Free energy is used to mask this inharmonious. But mass energy conversion is a greater challenge to the concept of "heat"! The Higgs mechanism uses the "vacuum" ground state energy to recover the mass of the particles. More generally, the space in which there are motions should have energy. All TS_B potential energy are determined by a number (number of states). And T is only the energy unit measuring this kinetic energy. This energy, which is most general and most relevant to heat, is likely to unify all kinds of force.

The present book has clarified and raised some of the problems: 1. Change of the TS_B potential energy corresponds to the heat exchanged between the system and the environment. No matter whether the original momentum type is changed or not, the thermal effect will appear as long as the original momentum spatial configuration changes; 2. The process creating heat can be done with a mass carrier, but the heat itself has no mass.

Since it is changed motion distribution. But the TS_B potential energy can have mass carrier or no mass carrier. But a kind of mass carrier can participate in several movements with different momentum at the same time. 3. Heat does not annihilate in the environment. Each momentum has its ground state space, in which the ground state energy of this motion should be accumulated when the momentum disappears.4. $\Delta(TS_B)$ of the spontaneous symmetry breaking process should point to the time arrow. Heat is an energy accumulation in the time sequence.

The TS_B potential energy can help us to obtain universal 'heat' concept.

To put it bluntly, TS_B energy's ability to construct a momentum field may just be the demiurge of an orderly world.

REFERENCES

[1] Xiaohua, W., Rarviz, M. (2009). Forest of hairpins in a low-reynolias-number zero-pressure-gradient flat-plate boundary layer. *Physics of Fluids*, 21. 091106.

[2] Oks, D., Mininni, P.D., Marino, R., Pouquet, A. (2017). Inverse cascades and resonant triads in rotating and stratified turbulence, *Physics of Fluids,* 29. 111109.

[3] Kraichnan, R.H. (1966). Isotropic turbulence and inertial-range structure, *Physics of Fluids,* 9.1728.

[4] Weimin, Huang, Jingxin, Gen, Xuping, Jan (1992). Laminar Taylor vortex structure and attractors, *International Conference on Fluid Mechanics and Theoretical Physics in honor of professor Pei-Yuan Zhou's 90th anniversary*, Abstracts 63.

[5] Jinfa, D., Kangnian, F. (1992). *Physical Chemistry,* Higher Education Press, Beijing (in Chinese).

[6] Pei, Z., Di, L. (2018). Catching the Birth of a dark molecular cloud for the first time, *The Astro Physical Journal,* 867.

[7] Stevenson, J., Lunine, J. & Clancy, P. (2015). Membrane alternative in worlds without oxygen of an azotosome, *Science Advances,* 1(1), e1400067.

[8] Junfeng, Z. (2019). Laboratory formation and photo-chemistry of ionic HBC/anthracene clusters in the gas phase, *Monthly Notices of the Royal Astronomical Society,* 486(3): 3259-3265.

[9] Philips, Rob, Kondev, Jane, Theriot, Julie (2009). Physical Biology of the Cell, *Garland Science,* (2012) Science Press (in Chinese Traslation Edition).

[10] Department of Physical Chemistry, Tianjin University, (2001). *Physical Chemistry,* Higher Education Press, Beijing (in Chenese).

[11] Han, Degang, Gao, Jidi, Gao, Panliang (2001). Physical Chemistry, *Higher Education Press,* Beijing, (in Chenese).

[12] Philips, Rob, Kondev, Jane, Theriot, Julie (2009). Physical Biology of the Cell, *Garland Science,* (2012) Science Press (in Chinese Traslation Edition)

[13] Zhang, J., Chen, P.C., Yuan, B.K., Ji, W., Cheng, Z.H., Qiu, X.H. Real-Space Identification of intermolecular Bonding with Atomic Force Microscopic, *Science*, (2013), 342 (6158),611-614.

[14] Philips, Rob, Kondev, Jane, Theriot, Julie (2009). Physical Biology of the Cell, *Garland Science,* (2012) Science Press (in Chinese Traslation Edition).

[15] Roger, Dodd., John, Gibbon,. Hedley, Morris, (1982). *Solitons and Nonlinear Wave Equations,* Academic Press, 14-15.

ABOUT THE AUTHOR

Weimin Huang has a BS degree in polymolecule. He worked in the industry as an engineer for decades. Professor Huang got a master's degree in hydrodynamics, heat, and mass transfer, and became a professor at the University of Shanghai for science and technology. The Chinese Science Fund Commission provides long-term support for Professor Huang's research projects. He has tracked, learned and explored thermodynamics for a long time in his teaching activities. The "Maxwell's demon" problem has been included in Professor Huang's study of "free energy" from the last twenty years, and gave him the chance to write this book.

INDEX

A

B

C

D

E

F

G

H

I

J

L

M

N

O

P

Q

R

S

T

V

W